Smart Nanosystems for Biomedicine, Optoelectronics and Catalysis

Edited by Tatyana Shabatina and Vladimir Bochenkov

Published in London, United Kingdom

IntechOpen

Supporting open minds since 2005

Smart Nanosystems for Biomedicine, Optoelectronics and Catalysis
http://dx.doi.org/10.5772/intechopen.83226
Edited by Tatyana Shabatina and Vladimir Bochenkov

Contributors
Fuwei Zhuge, Yuqian Chen, Yu Huang, Hongcheng Ruan, Bipin Deochand Lade, Julien Bras, Le Gars Manon, Lorelei Douard, Naceur Belgacem, Mondher Yahya, Faouzi Hosni, Ahmed Hichem Hamzaoui, Daniel Bellet, Dorina T. Papanastasiou, Joao Resende, Viet Huong Nguyen, Carmen Jiménez, Ngoc Duy Nguyen, David Muñoz-Rojas, Alessandro Fanti, Matteo Bruno Lodi, Tatyana Shabatina, Vladimir Bochenkov, Liangxing Hu, Nan Wang, Kai Tao, Gülay Baysal, Zihni Onur Onur Uygun, Hilmiye Deniz Ertuğrul Uygun, Arti S. Shanware

First published in London, United Kingdom, 2020 by IntechOpen
IntechOpen is the global imprint of INTECHOPEN LIMITED, registered in England and Wales, registration number: 11086078, 5 Princes Gate Court, London, SW7 2QJ, United Kingdom
Printed in Croatia

British Library Cataloguing-in-Publication Data
A catalogue record for this book is available from the British Library

Additional hard and PDF copies can be obtained from orders@intechopen.com

Smart Nanosystems for Biomedicine, Optoelectronics and Catalysis
Edited by Tatyana Shabatina and Vladimir Bochenkov
p. cm.
Print ISBN 978-1-83880-253-0
Online ISBN 978-1-83880-254-7
eBook (PDF) ISBN 978-1-83968-407-4

For EU product safety concerns:
IN TECH d.o.o., Prolaz Marije Krucifikse Kozulić 3, 51000 Rijeka, Croatia, info@intechopen.com or visit our website at intechopen.com.

Meet the editors

Prof., Dr. Tatyana I. Shabatina is the Head of the Laboratory on Low Temperature Chemistry of the Department of Chemistry, M.V. Lomonosov Moscow State University. She graduated in Chemistry with honors in 1978 (MSU), 1984 - Ph.D. in Physical Chemistry (MSU), 2013 - Doctor of Chemical Sciences, in Physical Chemistry (MSU), 1994 - research training in Max-Plank Institute, Muelheim (Germany), 1996 -research training at the University of Amsterdam (Nederland), 2000 - research scientist in the Kansas State University (USA), 2009 - Visiting Professor, exchange visit with the University of York (UK). Her scientific activity is connected mainly with cryo- and nanochemistry, metal nanoclusters and hybrid metal-mesogenic nanosystems, drug nanoforms and cryospectroscopy. Her awards include the M.V. Lomonosov Moscow State University Prize for young scientists (1984) and the Diploma of winner of M.V. Lomonosov Moscow State University Innovation Projects Exhibitions (2004, 2012, 2016). She was the supervisor of 8 PhD students and 12 undergraduates. She is the author of more than 120 scientific papers, 4 books and 4 book chapters, and 6 patents of Russian Federation.

Dr. Vladimir E. Bochenkov is a Leading Scientist of Lomonosov Moscow State University (MSU). He achieved his M.Sci. in Chemistry (2000) and PhD in Physics and Mathematics (2004) from MSU. Since 2004 he has been working as a scientific researcher at MSU. From 2011 to 2016, he has been working as a Postdoctoral Researcher at the iNANO Center of Aarhus University. His scientific activity has been connected mainly with nanomaterials, including the development of new nanofabrication techniques, experimental studies of plasmonic nanomaterials combined with computer simulation of their properties, and molecular dynamics simulations of stimuli-responsive polymers. He is an author of 41 scientific publications, 1 book, and 1 book chapter.

Contents

Preface

The book describes modern trends in nanoscience and nanotechnology for creation of smart hybrid nanosystems combining inorganic nano-objects with organic, biological, and biocompatible materials, which create multifunctional and remotely controlled platforms for diverse technical and biomedical uses. The published material includes several review and original research articles devoted to the problems of directed chemical and biological synthesis of such nanosystems, thorough analysis of their physical and chemical properties and prospects of their possible applications.

The combination of magnetic nanoparticles and biocompatible materials leads to the manufacturing of a multifunctional and remotely-controlled platform useful for diverse biomedical applications. The first chapter, "Biomedical Applications of Biomaterials Functionalized with Magnetic Nanoparticles" by Matteo B. Lodi and Alessandro Fanti, describes the possibilities of the formation of such hybrid nanosystems and the variety of their biomedical use. It covers the questions of mathematical modeling of the drug delivery processes and assesses the problem of establishing the influence of the system on tissue regeneration. On the other hand, if a time-varying magnetic field is applied, the magnetic nanoparticles would dissipate heat, which can be exploited to perform local hyperthermia treatment on residual cancer cells in bone tissue. To perform the treatment planning, it is necessary to account for the modeling of the intrinsic non-linear nature of the heat dissipation dynamics in magnetic prosthetic implants. In this work, numerical experiments to investigate the physio-pathological features of the biological system, linked to the properties of the nanocomposite magnetic material, to assess its effectiveness as therapeutic agent are presented.

Magnetic nanoparticles (MNPs) display physical and chemical properties different from those found in their corresponding bulk materials. These properties make them attractive in various applications such as energy, electronics, sensor designs of all kinds, catalysts, magnetic refrigeration, optics, and in various biomedical applications. These multifunctional nanomaterials can be used as contrast agents for medical imaging, nano-vectors to transport therapeutic agents to their target, local delivery of drugs or used to destroy the cancer cells by local hyperthermia. These magnetic platforms should possess small size combined with high magnetic susceptibility and loss of magnetization after removal of the magnetic field. Chapter 2, "Synthesis and ESR Study of Transition from Ferromagnetism to Superparamagnetism in $La_{0.8}Sr_{0.2}MnO_3$ Nanomanganite" by Mondher Yahya, Faouzi Hosni, and Ahmed Hichem Hamzaoui, describes the questions of the optimization of the nanoparticle's size, size distribution, agglomeration, surface coatings and shapes along with the changes in magnetic properties prompted by the application of magnetic nanoparticles in diverse fields.

Nowadays, plasmonic nanostructures attract an increasing attention as signal amplifiers and transducers for optical sensing. The local plasmon-induced enhancement of electric fields affects various optical processes in molecular systems and therefore finds various applications in enhanced spectroscopic techniques, such as Surface-Enhanced Raman Scattering (SERS), Plasmon-Enhanced Fluorescence (PEF), Surface-Enhanced Infrared Absorption (SEIRA), etc. Chapter 3, "Chiral

Hybrid Nanosystems and Their Biosensing Applications" by Vladimir E. Bochenkov and Tatyana I. Shabatina, is devoted to chiral biosensing using various metal-containing hybrid nanosystems based on optically active organic and biological molecules. Plasmonic nanosystems and nanostructures provide an excellent platform for label-free detection of molecular adsorption by detecting tiny changes in the local refractive index or by amplification of light-induced processes in biomolecules. Based on recent theoretical and experimental developments in plasmon-enhanced techniques, we consider the main types of plasmonic nanosystems capable of generating an amplified chiroptical signal for such applications as detecting the presence of certain biomolecules and, in some cases, for the determination of molecular orientations and their higher-order supramolecular structure.

Sensor and biosensor technologies have demonstrated rapid progress in recent years. These technologies use nanosystems that are highly important in immobilization of materials for recognition the target molecules. Chapter 4, "Fullerene Based Sensor and Biosensor Technologies" by Hilmiye Deniz Ertuğrul Uygun and Zihni Onur Uygun, describes a number of studies of fullerene-based sensor nanomaterials. As zero-dimensional nanomaterials, fullerenes provide an extremely large surface area. Therefore, they provide more biological or non-biological recognition receptors immobilized on this surface area. Moreover, increasing the surface area with more recognition agent also increases the sensitivity. In this book chapter, the examples of the development of fullerene-based sensor and biosensor technologies and their modifications and the comparison of fullerene-type sensor and biosensor applications in different cases are presented and discussed.

Carbon nanomaterials can increase the sensitivity of different diagnostic systems due to the increase of surface area and conductivity. Chapter 5, "Cellulose Nanocrystals: From Classic Hydrolysis to the Use of Deep Eutectic Solvents", by Manon Le Gars, Loreleï Douard, Naceur Belgacem, and Julien Bras, describes recent achievements in the development of a promising sub-branch of diagnostic systems–point-of-care diagnostic tests–which has made great progress due to the use of the fundamentals of sensor and biosensor nanotechnology. The synthesis of cellulose nanocrystals for diagnostic systems development is considered in detail.

The greener way of producing metal nanoparticles is the easiest, cheapest, and the most efficient way of producing large-scale nanoparticles that have no adverse effect on the environment. Chapter 6, "Phytonanofabrication: Methodology and Factor Affecting Biosynthesis of Nanoparticles" by Bipin D. Lade and Arti S. Shanware, discussed in detail the preparation of silver nanoparticles using various methodologies and the biological synthesis. The effects of various sources and methods on nanoparticle synthesis, the impact of conditions such as dark, light, heating, boiling, sonication, autoclave on the size and shape of colloidal nanoparticles have been analyzed. The authors discuss the effects of specific parameters such as leaf extract concentration, $AgNO_3$, reaction temperature, pH, light, and stirring time for nanoparticle synthesis and the results on the impact of silver nanoparticles on plant physiology.

Smart nanosystems are used in many fields such as medicine, biomedical, biotechnology, agriculture, environmental pollution control, cosmetics, optics, health, food, energy, textiles, automotive, communication technologies, agriculture, and electronics. Chapter 7, "The Components of Functional Nanosystems and Nanostructures" written by Gülay Baysal, is devoted to the questions of new synthetic, diagnostic, and treatment methods for production of nanospheres, nanorobots, biosensors, quantum dots, and biochips, which are considered to be the main components of smart nanosystems and the combination of nanotechnological approaches based on mimicking the principle of atomic sequence in nature for their creation.

Chapter 8, "Metallic Nanowire Percolating Network: From Main Properties to Applications" by Daniel Bellet, Dorina T. Papanastasiou, Joao Resende, Viet Huong Nguyen, Carmen Jiménez, Ngoc Duy Nguyen, and David Muñoz-Rojas, is devoted to metallic nanowire (MNW) networks that appear to be one of the most promising flexible, efficient, and low-cost transparent electrodes that can be integrated for many applications. This includes several applications related to energy technologies (photovoltaics, lighting, supercapacitor, electro-chromism) or displays (touch screens, transparent heaters) as well as Internet of Things linked with renewable energy and autonomous devices. Randomly deposited MNW - AgNW or CuNW networks present record values of sheet resistance values below 10 Ω/sq, optical transparency of 90% and high mechanical stability under bending tests. Networks are destined to address a large variety of emerging applications. The main properties of MNW networks, their stability, and their integration in energy devices are discussed in this contribution.

In Chapter 9, "Emerging Artificial Two-Dimensional van der Waals Heterostructures for Optoelectronics" written by Hongcheng Ruan, Yu Huang, Yuqian Chen and Fuwei Zhuge, the authors introduce the basic concept of the design of hybrid nanosystems and hetero-nanostructures for optoelectronics and describe the pick-transfer methods for their artificial assembly. They discuss the recent progress in fabricating novel 2D van der Waals heterostructures for functional devices. In view of the rapid progress in this field, the chapter is not intended to cover all aspects of the field but focuses on optoelectronic related application, typically photodiode and phototransistors for photodetection and optoelectronic memories that integrate both light sensing and memory function.

Chapter 10, "Catalytic Micro/Nanomotors: Propulsion Mechanisms, Fabrication, Control and Applications" by Liangxing Hu, Nan Wang and Kai Tao, describes self-propelled micro/nanomachines and micro/nanomotors, which are capable of converting the surrounding fuels into mechanical movement or force. Inspired by naturally occurring biomolecular motor proteins, scientists extensively paid great attention to synthetic micro/nanomotors. Especially, the authors describe the possibility of the creation of catalytic micro/nanomotors. The future of this research field can be bright, but some major challenges such as biocompatible materials and fuels, smart controlling, and specifically practical applications still need to be resolved. In this chapter, propulsion mechanisms, fabrication methods, controlling strategies, and potential applications of catalytic micro/nanomotors are presented and summarized.

We hope that this book will be useful for different nanoscience research groups and PhD and graduate students, to introduce them to the world of hybrid metal-organic and metal-biological nano-objects and smart self-organizing nanosystems and open new views of possible use of them in different scientific and practical areas.

Tatyana I. Shabatina and Vladimir E. Bochenkov
Lomonosov Moscow State University,
Moscow, Russia

Chapter 1

Biomedical Applications of Biomaterials Functionalized with Magnetic Nanoparticles

Matteo Bruno Lodi and Alessandro Fanti

Abstract

The combination of magnetic nanoparticles and a biocompatible material leads to the manufacturing of a multifunctional and remotely controlled platform useful for diverse biomedical issues. If a static magnetic field is applied, a magnetic scaffold behaves like an attraction platform for magnetic carriers of growth factors, thus being a potential tool to enhance magnetic drug delivery in regenerative medicine. To translate in practice this potential application, a careful and critical description of the physics and the influence parameter is required. This chapter covers the mathematical modeling of the process and assesses the problem of establishing the influence of the drug delivery system on tissue regeneration. On the other hand, if a time-varying magnetic field is applied, the magnetic nanoparticles would dissipate heat, which can be exploited to perform local hyperthermia treatment on residual cancer cells in the bone tissue. To perform the treatment planning, it is necessary to account for the modeling of the intrinsic nonlinear nature of the heat dissipation dynamic in magnetic prosthetic implants. In this work, numeric experiments to investigate the physiopathological features of the biological system, linked to the properties of the nanocomposite magnetic material, to assess its effectiveness as therapeutic agents are presented.

Keywords: biomaterials, bone tumors, bone repair, drug delivery, hyperthermia, magnetic nanoparticles, RF heating, scaffolds

1. Introduction

Nanotechnologies aim to ease and to satisfy the needs of regenerative medicine[1] by providing multifunctional, theranostic, and stimuli-responsive biomaterials [1, 2]. In particular, stimuli-responsive biomaterials such as magneto-responsive biomaterials are devices capable of reacting to an external magnetic field spatio-temporally in a specific way [3]. This powerful class of biomaterials is a promising candidate as active and therapeutic scaffolds for advanced drug delivery and tissue regeneration applications [3, 4].

Multifunctional magnetic-responsive materials can be manufactured by modifying or functionalizing traditional materials employed in tissue engineering or by

[1] Regenerative medicine is a tissue regeneration technique based on the replacement or repair of diseased tissue or organs to restore a lost or impaired function [1].

Type of scaffold	Synthesis technique	M_s, emu·g^{-1}	Type of MNPs	r_{mnp}
HA/collagen	Impregnation	0.35–15	Fe_3O_4	200
HA/collagen	Impregnation	0.50	γ-Fe_2O_3, Fe_3O_4	10–50
HA/PLA	Electrospinning	0.05	γ-Fe_2O_3	5
β-TCP	Impregnation	0.6–1.2	Fe_3O_4	250
Chitosan/PVA membrane	Electrospinning	0.7–3.2	Fe_3O_4	n.s.
Calcium silicate/chitosan	Mixture	6–10	$SrFe_{12}O_{19}$	500
PMMA	Mixture	n.s.	Fe_3O_4	10
Silicate	Mixture	n.s.	γ-Fe_2O_3	n.s.
Fe-doped HA	Chemical substitution	4	HA-Fe_3O_4	10–14
Fe-hardystonite	Chemical doping	0.1–1.2	Fe_3O_4	20–60
Bredigite	Milling	7–25	$Ca_7MgSi_4O_{16}$-Fe_3O_4	120
HA	Impregnation	12–20	Fe_3O_4	200
HA	Impregnation	1–2.5	γ-Fe_2O_3	8
HA	Impregnation	n.s.	γ-Fe_2O_3	5
Chitosan	In situ precipitation	4	γ-Fe_2O_3, Fe_3O_4	n.s
ϵ-PCL	3D Bioplotting	0.2–0.3	Fe_3O_4	25–30
PLGA	Electrospinning	2–10	Fe_3O_4	8.47

Table 1.
Magnetic scaffolds divided by composition, production, and MNPs embedded. Redrafted from [5].

incorporating magnetic nanoparticles (MNPs) in the biocompatible matrix [4, 5]. **Table 1** reports examples of several magnetic biomaterials synthesized in the literature [6]. An approach to create a magnetic biomaterial is the impregnation of a polymer or ceramic (e.g., ϵ-poly caprolactone or hydroxyapatite) with MNPs dispersed in a ferrofluid (FF) [5, 6]. Subject to the action of capillarity, the nanoparticles fill the superficial defects and pores of the biomaterials. In this way a nanocomposite is created, i.e., the final material is a two-phase system strengthened by the magnetic iron phase [7]. Moreover, a multifunctional and composite material of such type can be obtained by the polymerization of a polymer in the presence of magnetic nanoparticles of magnetite (Fe_3O_4) or maghemite (γ-Fe_2O_3). This allows to produce a solid object using rapid prototyping and additive manufacturing techniques, such as electrospinning or 3D bioplotting [7].

In alternative, a stable, repeatable, and controllable manufacturing technique of magnetic-responsive biomaterial is the chemical doping of or substitution with F^{2+} or F^{3+} ions in a ceramic material (e.g., hydroxyapatite, β-tricalcium phosphate, and hardystonite). This process gives rise to an intrinsic magnetic and biocompatible material, which can be used in the form of microparticles or directly as a bulk object with tunable and ad hoc properties for therapeutic or regenerative medicine applications [8, 9].

Given these methods, the magnetic biomaterial can be processed to develop a tissue-guiding structure or a tissue scaffold, i.e., a device intended to be implanted in an injured site for supporting and withstanding the cell adhesion, proliferation, and differentiation, in order to restore tissue continuity and functioning [10]. Magnetic scaffolds (MagS) have been proposed for the following three main applications, as presented in **Figure 1** [1–9]:

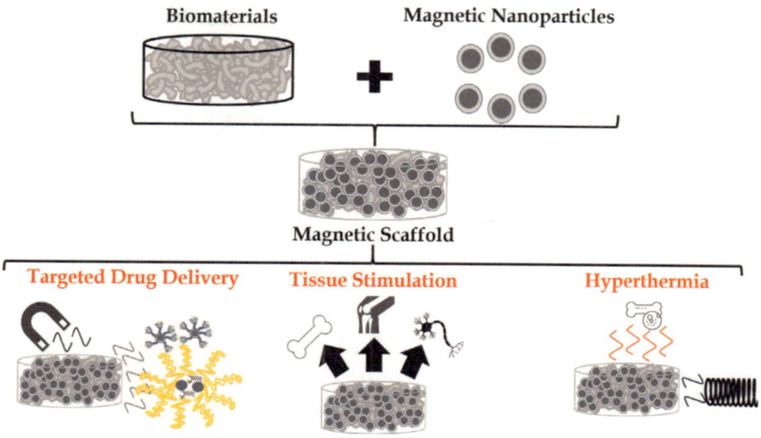

Figure 1.
Magnetic scaffolds are obtained by the combination of biomaterials and MNPs. They are multifunctional and theranostic nanocomposites. The potential biomedical applications of MagS are shown.

- To provide a controlled *mechanical stimulation* of tissues and boost the healing response
- To develop a smart and reliable *magnetic drug delivery system* (MDD)
- To generate therapeutic heat and perform local *hyperthermia* (HT) against cancer cells

The mechanical stimulation of injured tissues using magneto-responsive scaffolds found application in bone tissue engineering, where static magnetic field (SMF) or low-frequency magnetic field is used to elicit osteoprogenitor cells [1–4].

The rationale of employing magnetic scaffolds as part of a MDD system is the need to have an "attraction platform" to target and control the attraction of magnetic liposomes or MNPs bio-conjugated with growth factors (GFs) [6, 11]. Indeed, recently several magnetic carriers of biomolecules capable of acting on cell function were developed. However, using an external SMF their delivery to deep tissue and to the site of damage is complicated, and the MNPs tend to distribute where the magnetic force is maximum, i.e., at the body surface, where the field is applied [12]. Having a MagS implanted in the injured tissue allows to attract the MNPs and the GFs while controlling their spatial distribution [13].

Finally, if the external magnetic stimulus is a radio-frequency (RF) magnetic field, the population of MNPs embedded in the biomaterial dissipates a huge amount of heat. The deposited power can be exploited as therapeutic heat, enabling to use the magnetic scaffold as a thermo-seed able to perform HT treatment against cancer cells [14].

To date, magnetic scaffolds have been synthesized and characterized in terms of chemical and physical properties while proving experimentally their powerful and promising potential in regenerative medicine and oncology [1–4]. However, to translate the use of these nanostructured biomaterials in the clinical practice, several limitations have to be overcome, and further investigations are required to predict their behavior [4]. The potential use of magnetic scaffolds as tissue substitutes needs the combined work of material scientists, biomedical engineers, and biologists. In particular, since in the literature there is a clear lack of mathematical and numerical models, which relate the physical properties of these nanocomposite biomaterials with the magnetic drug delivery or the hyperthermia, in this chapter, two mathematical models for their use as hyperthermia agent and as a tool for magnetic drug delivery are provided.

Section 2 briefly reviews the use of MagS as magneto-responsive biomaterials for the stimulation of tissues, in particular bone tissues. In Section 3 the nonlinear chemico-physical properties of magnetic scaffolds are presented, described, and used to introduce a recent in silico model for the planning of bone tumor hyperthermia [14]. Finally, in Section 4 the use of MagS as tool for active magnetic drug delivery is discussed. Furthermore, a mathematical model able of providing insights into the parameters of influence of the phenomenon is presented and analyzed [13]. The complete description of magnetic scaffolds favors the assessment of their effectiveness and their potential clinical impact.

2. Magnetic scaffolds for tissue repair and regeneration

Magnetic scaffolds have been tested both in vitro and in vivo, using animal models, demonstrating that they can transduce an external magnetic signal in mechanical stimulation to the cells attached to the biomaterial surface (**Figure 1**) [1–4]. MagS have been investigated for bone, cartilage, cardiovascular and neuronal regeneration, and repair [2]. The most studied tissue is bone. The injury of skeletal tissue by traumas and diseases, such as osteoporosis, or by a tumor resection calls for the need of a bone substitute or scaffold to guide cell adhesion, proliferation, and differentiation [15]. Moreover, the bone tissue requires a continuous mechanical stimulation. Therefore, the magneto-responsive biomaterials in **Table 1** can deliver a direct mechanical stimulation if exposed to SMF, to low-frequency magnetic field (strengths from to 18 μT to 0.6 T, frequencies varying from 10 to 76.6 Hz), or to pulsed electromagnetic fields [4]. The mechanism of action is not fully understood yet. The presence of magnetic nanoparticles in the biomaterials determines an increased superficial roughness and favors the interaction at the cell membrane with the cell surface receptors. It has been demonstrated that the mesenchymal stem cells (MSCs) can differentiate into osteoblast thanks to the activation of the integrin signaling pathways, which upregulate the expression of the osteogenic GF bone morphogenetic protein 2 (BMP-2) [4]. The use of magnetic scaffolds permits the integration of the implant with the host tissue, accelerating the defect healing and increasing the mineral density of newly formed bone.

3. The hyperthermia treatment of bone tumors

3.1 The heat dissipation of magnetic nanoparticles

To understand the magnetization dynamic and the power losses of magnetic scaffolds, it is necessary to introduce the physical and mathematic descriptions of the response to a RF magnetic field of the MNPs embedded in it. If a population of magnetic nanoparticles in a solution is exposed to a harmonic RF magnetic field, they start to dissipate power due to the hysteresis loss but also to the magnetic dipole and to the Brownian relaxations [16]:

$$P_m = \pi\mu_0 f|\underline{H}|^2\chi'' \tag{1}$$

where μ_0 is the vacuum permeability; f is the frequency of the applied field, in Hz; $\overline{H}$ is the amplitude of the external magnetic field; and χ'' is the imaginary part of the complex magnetic susceptibility of the particles. For ferrofluids, the magnetic susceptibility is known to be described by the Debye model [13, 16]:

$$\chi(f) = \chi' - j\chi'' = \frac{\chi_0}{1 + j2\pi f\tau} \tag{2}$$

The term χ_0 is the equilibrium susceptibility that is defined as [17]:

$$\chi_0 = \chi_i \frac{3}{\zeta}\left(coth\ (\zeta) - \frac{1}{\zeta}\right) \tag{3}$$

where ζ is the ratio between the magnetic energy of the set of magnetic dipoles and the thermal energy. Mathematically speaking:

$$\zeta(T) = \frac{\mu_0 \phi M_s^2 V_{mnp} |\overline{H}|}{3kBT} \tag{4}$$

where M_s is the saturation magnetization of the single MNPs, in Am^{-1}; V_{mnp} is the particle volume in nm^3; k_B is the Boltzmann's constant; and T is system temperature. In Eq. (3), χ_0 is the initial susceptibility, which is defined as [17]:

$$\chi_i(\overline{H}, T) = \frac{\mu_0 \phi M_s V_m |\overline{H}|}{3kBT} \tag{5}$$

The term τ in the Debye model (Eq. 2) is the effective relaxation time, in s, which can be evaluated as the parallel of the Néel and Brownian processes [17]:

$$\frac{1}{\tau} = \frac{1}{\tau_N} + \frac{1}{\tau_B} \tag{6}$$

The time required to the magnetic dipole moment to align with the direction of the external magnetic field is called the Néel relaxation time, τ_N [16, 17]:

$$\tau_N = \frac{\sqrt{\pi}}{2} \tau_0 \frac{e^{\Gamma}}{\Gamma} \tag{7}$$

The pre-exponential factor τ_0 is a time, and its value can range from 0.1 ps to 1 ns, but this term is a function of system temperature, i.e., $\tau_0 = \tau_0(T)$ [13]. The term Γ is the ratio of the anisotropy energy of the nanoparticle to the thermal energy of the system, i.e.:

$$\Gamma = \frac{K_a V_m}{k_B T} \tag{8}$$

where K_a is the magnetic anisotropy energy per unit volume in Jm^{-3} and V_m is the MNP volume in nm^3.

In a FF, the nanoparticles are allowed to rotate and move according to Brownian motion in the viscous medium where they are dispersed. When subject to a time-varying magnetic field, the particles rotate to orient with the direction of the external energy source, thus contributing to the relaxation process. The Brownian relaxation time can be evaluated as [16]:

$$\tau_B(\eta, T) = \frac{3\eta V_h}{k_B T} \tag{9}$$

being η the viscosity of the medium, in Pa·s, and V_h the hydrodynamic radius of the particle in solution.

With Eqs. (1) to (9), it is possible to describe the frequency response and the power dissipation of a population of MNPs dispersed in a solution. This set of equations constitutes the theoretical basis for the understanding of magnetic scaffold behavior. However, since MagS are solid nanocomposites, the behavior of their magnetic phase is rather diverse than a FF. In the following, the experimental findings related to material characterizations and a new mathematical framework to account for their response are provided.

3.2 Hyperthermia response of magnetic scaffolds

Hyperthermia (HT) is a thermotherapy which aims at increasing the temperature of a target tissue between 41 and 46 C for about 60 min. For biological tissues, especially neoplasms and cancers, these temperatures are sufficient to damage the DNA of cells, altering its replication and also the repair pathways while determining cytotoxicity and activating the response of the host immune system [18, 19]. The rather chaotic vascular architecture of tumors is the reason of the thermo-sensibility of these pathologic tissues. The aforementioned biological effects can lead to the death of cancer cells, but, in the clinical practice, HT is exploited as a coadjuvant therapy combined with chemotherapy or/and radiotherapy rather than as a standalone therapy [19]. The hyperthermia can be induced using different types of energies, such as ultrasounds or electromagnetic (EM) field [14]. Currently different therapeutic modalities are available for HT induced by EM field. In particular, it is thoroughly investigated the local and in situ treatments using nanoparticles or magnetic scaffolds by exposing the target are with an external magnetic field.

Several magnetic scaffolds from **Table 1** demonstrated to be capable of noticeable temperature increases when exposed to magnetic field working at the frequencies from 100 kHz to 1 MHz and with amplitude ranging from 8 to 25 kAm^{-1} [5, 7]. Different biomaterials (e.g., hydroxyapatite, β-TCP, and PCL) loaded with magnetite or maghemite nanoparticle temperature increase, from room temperature, of 8–45°C, were measured [5]. In particular, on one hand, a scaffold made with the intrinsic magnetic hydroxyapatite of Tampieri et al. [8] can increase the temperature of 40°C in 60 s when exposed to a magnetic flux density with amplitude of 30 mT and frequency 293 kHz. On the other hand, in the same exposure conditions, a PCL scaffold loaded with magnetite nanoparticles can raise the temperature from 20 to 32°C in 600 s [7, 8, 13].

These composite nanomaterials are identified as optimal candidates for local bone tumor hyperthermia [1–9, 13, 14]. However, their therapeutic potential must be investigated in a critique way. The understanding and the modeling of the heat dissipation of the MNPs embedded in the biomaterial are essential to allow an effective treatment planning.

3.3 The susceptibility spectra of magnetic scaffolds

The physical explanation of the relevant and significant temperature increases measured for MagS is not trivial. Moving from the theory explained in Section 3.1, the resonant Debye model cannot be applied to a system in which highly concentrated MNPs are fixed and embedded in a solid matrix and lattice or constrained in a highly viscous medium [13]. Indeed, the long-range interactions between the magnetic nanoparticles become relevant [20]. The following index Υ, given in [20], can be considered to estimate the level of dipole–dipole interactions in the nanosystem under analysis:

$$\Upsilon = \frac{\mu_0 \mu_{mnp}^2}{4\pi k_B T d_{mnp}^3} \quad (10)$$

where the cubic power of the particle diameter, d_{mnp}^3, is the lower limit of the volume packaging and of the steric hindrance in the system [5]. For example, for magnetite nanoparticles the particle diameter measured in dry condition with transmission electron microscope (TEM) is lower than the hydrodynamic radius assessed using dynamic light scattering (DLS), i.e., 10 nm against 25 nm. Therefore, at the body temperature of 37°C, given the same dipole moment μ_{mnp}, the level of interaction of MNPs in solution is 2.5 times lower. It is demonstrated by the morphological and structural characterization of MagS that the MNPs in the biomaterial are often aggregated or very near, implying that small clusters of nanoparticles can be identified [5]. Moreover, this last evidence supports the theory for which the relaxation dynamic of clusters of MNPs is strongly modified due to the appearance of a distribution of anisotropy energies [20]. In other words, the Neel relaxation time in Eq. (7) will depend on the number and the dispersion of sizes of the nanoparticles. Another limitation of the applicability of the Debye model to the description of magnetic scaffolds is the influence of Brownian relaxation time on the heat dissipation mechanism. It has been demonstrated that the frequency response of the complex magnetic susceptibility $\chi(f)$ of MNPs modifies if the particles are constrained in agarose gel or used as cross-linkers in hydrogels [21, 22]. This change in the susceptibility spectra is due to the fact that in a highly viscous matrix or in a solid, the Brownian mechanism is hindered or canceled. From Eq. (9), in mathematical terms:

$$\lim_{\eta \to \infty} \tau_B(\eta, T) \to 0 \quad (11)$$

Therefore, in MagS the only relaxation time is the Néel one.

The influence of long-range interactions between particles, the modified distribution of anisotropy energy, and the different Néel relaxation dynamic are the factors that contribute to enhance the power dissipation of magnetic scaffolds, and all of them can help to explain the hyperthermia behavior of MagS, such as for the magnetic hydroxyapatite and the Fe-doped PCL scaffolds [7]. Relying on the magnetic susceptibility spectra of MNPs in agarose gel measured by Hergt et al. [21], a Cole-Cole model for magnetic scaffolds [13]:

$$\chi(\omega) = \frac{\chi_0}{1 + (2\pi f \tau)^{1-\alpha}} \quad (12)$$

Equation (12) can fit the susceptibility data, with a 1.5% relative error, as shown in **Figure 2**, whereas the Debye model cannot (Eq. (2)). In Eq. (12) γ is the broadening parameter, which is found to be equal to 0.75 [13]. The differences in the frequency response are depicted in **Figure 2**.

With Eqs. (1)–(8), but using Eq. (12) instead of Eq. (2), it is possible to evaluate and estimate the power losses of magnetic scaffolds. At this point it should be noted that the magnetic susceptibility $\chi(f)$ is a function of the system temperature, which influences the initial susceptibility and the Néel relaxation time (Eqs. (5) and (7)). As the temperature increases, the therm τ_0 increases, whereas the time τ_N decreases. The outcome is a decrease in the imaginary part of the magnetic susceptibility, χ'', which in turn lowers the power deposited by the magnetic scaffold. Therefore, since the goal of hyperthermia treatment is to increase the temperature

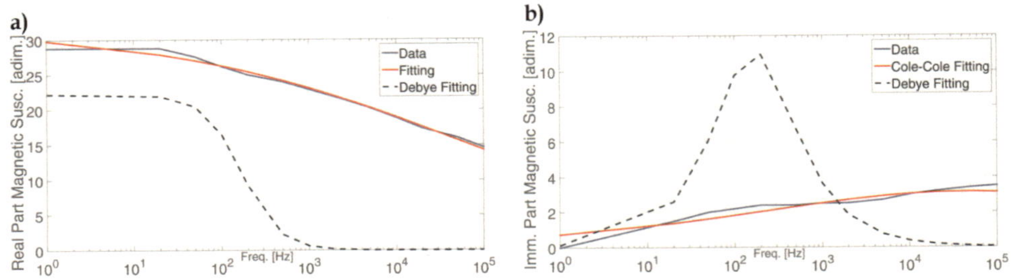

Figure 2.
Results of the fitting of the magnetic susceptibility spectra of MNPs embedded in agarose: a) real part (in-phase) and b) imaginary (out-of-phase) components are presented [21]. The Debye and Cole-Cole models are used and compared Taken from [13].

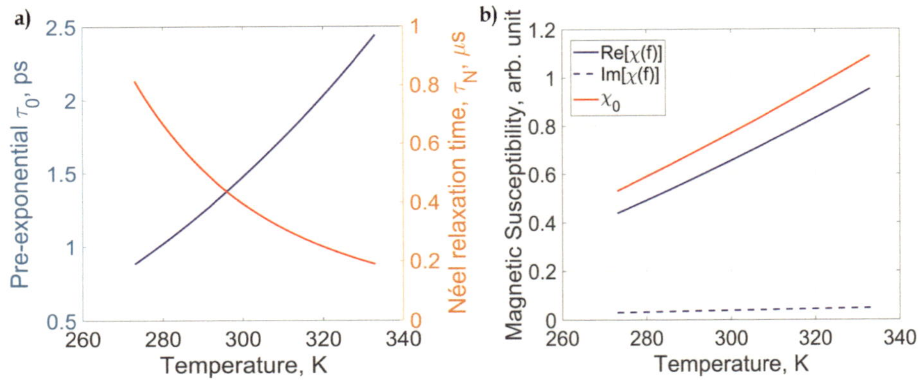

Figure 3.
Temperature variation of the pre-exponential term τ_0 and the Neel relaxation time τ_N. The influence on the equilibrium and the complex magnetic susceptibility χ_0 and $\chi(f)$ is represented. The curves are obtained for a magnetic scaffold filled with the 0.2% of magnetite nanoparticles (r_{mnp}=10 nm, $M_s(0)$ = 2 emu·g^{-1}, T_b=150 K).

of cancer tissues, it follows that the magnetic properties of and hence the power dissipated by MagS change during the treatment. The influence of temperature on the different physical quantities is shown in **Figure 3**. Since P=$P(T)$, planning a HT treatment which employs MagS as thermo-seeds against tumors is a multiphysics and highly nonlinear problem [14].

3.4 The hyperthermia treatment of bone tumors

Given the potential of magnetic scaffolds to be used as local heat source for setting the hyperthermia treatment of cancers, the most studied biological and clinical target of the nanosystems under investigation are bone cancers. Indeed, in clinical practice, currently available techniques such as chemotherapy, radiotherapy, and osteotomies presented a 15% probability of tumor recurrence, and therefore the hyperthermia treatment was proposed as adjuvant therapy [23]. Furthermore, since the surgical intervention causes a bone damage which calls for a graft or bone substitutes, magnetic scaffolds as theranostic, multifunctional, and magnetic-responsive biomaterials can be employed and can express their clinical potential [14].

Bone tumors are neoplasms mostly affecting subjects with age between 10 and 25 years old, causing impairment and pain, thus ruining the quality of life [24]. Malignant bone cancers such as osteosarcoma (OST) and fibrosarcomas (FIB) are known to affect long bone extremities [24]. OST and FIB are two different forms of bone cancer. The OST is big, aggressive and highly vascularized, whereas FIB is a poorly vascularized neoplasm. The survival rate for patients affected by OST and

FIB may vary from 28–40% [14, 23, 24]. To overcome these clinical issues, oncologist investigated the use of immunotherapy or smart nanocarriers of drugs, but local hyperthermia stands out as a very promising therapy [14]. The rationale is to implant a MagS after the bone tumor resection or reduction and then perform a local and in situ hyperthermia treatment by applying an external RF magnetic field. The residual cancer cells would be killed or increase their sensibility to drugs or radiations. Finally, the scaffolds would serve as supporting architecture for healthy cells, favoring tissue repair [14].

3.4.1 The in silico scenario

With the knowledge of the mechanism of power dissipation of MNPs embedded in a scaffold, recently a numerical scenario, with layered geometry, was proposed to investigate using finite element methods (FEM) the effectiveness of magnetic scaffolds in treating the residual bone cells of OST and FIB tumors [14].

As shown in **Figure 4**, imagining a surgical intervention of a bone cancer in distal femur, a spherical magnetic scaffold, with radius r_s = 5 mm, is implanted to fulfill the bone cavity [14]. A small gap or fracture (r_f = 0.1 mm) separates the scaffold from an annular region where the residual cancer cells of OST or FIB are supposed to be found. The fracture gap is a heterogeneous region where blood and bone are present, and it is where the new bone forms. It was demonstrated that its presence and biological status, i.e., if it is inflamed or ischemic, can influence the HT outcome [14]. The tumor area has a radius r_t which can vary from 0.1 to 0.5 mm. The goal of the HT treatment is to raise the temperature above 42°C for 30 min [14, 17]. Finally, a healthy bone tissue region with radius r_b=5 mm is included. The healthy bone should not be damaged by the treatment [5, 14]. It should be pointed out that muscle, fat, or skin tissues are not included in this analysis [25].

3.4.2 The electromagnetic problem

With respect to the geometry in **Figure 4**, the HT treatment using MagS is carried out applying an external RF magnetic field with strength H_0, working frequency f (293 kHz or 409 kHz [14]), and a time envelope able to keep the target temperature above the therapeutic threshold. The magnetic field is supposed to be

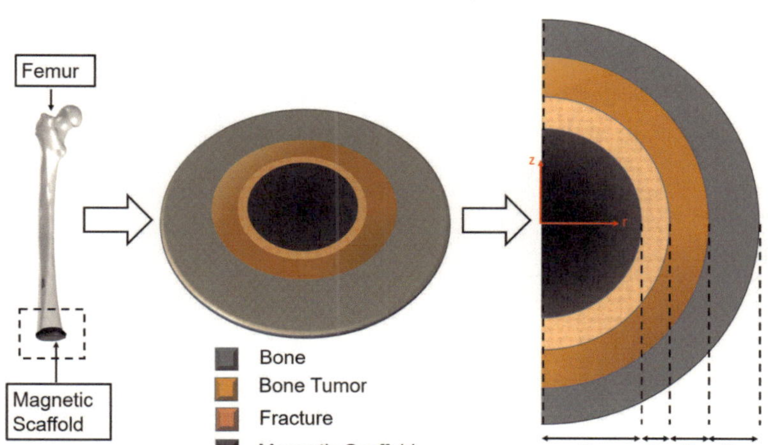

Figure 4.
Simplified layered geometry for modeling the hyperthermia treatment of bone tumors using magnetic scaffolds. The MagS with radius r_s = 5 mm is surrounded by a surgical fracture gap (r_f = 0.1 mm), the area where residual cancer cells are present (r_t = 0.1 mm–0.5 mm), and the healthy bone tissue (r_b = 5 mm). Taken from [14].

homogeneous in space [14]. At the initial time, the system is supposed to be in thermal equilibrium with a constant temperature distribution (T_0=37°C). Maxwell's equation in the frequency domain should be solved to calculate the power dissipated in the system [5]:

$$\begin{aligned} \nabla \times \overline{H} &= j\omega(\epsilon_0 \overline{E} + \overline{J}) \\ -\nabla \times \overline{E} &= -j\omega\mu_0 \overline{H} \end{aligned} \tag{13}$$

where ω is the angular frequency $2\pi f$, ϵ_0 is the vacuum permittivity in Fm^{-1}, $\overline{E}$ is the electric field vector in Vm^{-1}, and, finally, $\overline{J}$ conduction current vector in the system, in $A{\cdot}m^2$. The total electromagnetic power dissipated is the sum of the power density dissipated by the scaffold and the tissues, i.e., P_{EM} = P_m + P_e. The power deposited by the MagS, P_m, requires Eq. (13) to be solved considering the set of Eqs. (1)–(8) but using Eq. (12) instead of (2). The power dissipated by the induced currents in tissues (P_e) cannot be neglected, even though several mathematical models related to magnetic hyperthermia did not include it [26]. However, the dielectric losses in the system can have a significant contribution to the final temperature increase while causing the unwanted indirect heating of the nontarget tissues [5, 25].

The EM problem is solved employing the *RF module* of the commercial FEM software COMSOL Multiphysics (COMSOL Inc., Burlington, MA). The MagS studied are the intrinsic magnetic hydroxyapatite and the PCL loaded with magnetite [7], as in [14]. The dielectric properties of scaffold and tissues at T_0 are reported in **Table 2**.

3.4.3 The heat transfer problem

The power deposited by the MagS and conducted to the tissues in the system of **Figure 4** modifies the temperature ($T = T(r, z, t)$), whose spatiotemporal evolution can be evaluated using the Pennes' bioheat equation [14]:

$$\rho C_p \frac{\partial T}{\partial t} = \nabla \cdot (k \nabla T) + \rho_b C_{p,b} \omega_b (T - T_a) + Q_{met} + P_{EM} \tag{14}$$

where ρ is the density in $g{\cdot}m^{-3}$, C_p is the specific heat capacity in $Jkg^{-1}K^{-1}$, and k is the thermal conductivity in $Wm^{-2}K^{-1}$. The terms ρ_b and C_b are the density and heat capacity of blood, whereas the quantity ω_b is the tissue perfusion, in s^{-1}, i.e., the capillary contribution which acts to equilibrate the tissues with the blood temperature T_b=37°C. Q_M is the metabolic heat rate generated by the tissues, in Wm^{-3}.

Material or tissue	**Re[ϵ]**	**σ, Sm^{-1}**
Magnetic hydroxyapatite	12.5	$2.1{\cdot}10^{-3}$
ϵ-PCL	2.20	10^{-4}
Fracture gap–inflamed	3580	0.545
Fracture gap–ischemic	1321	0.196
Bone tumors: OST and FIB	8000	0.280
Bone	192	0.0214

Table 2.
Electromagnetic properties of scaffolds and tissues [14].

Eq. (14) was implemented in COMSOL using the *Bio-Heat transfer module*. The initial temperature T_0 was set in all domains. The bone edges are assumed to be open boundaries. The temperature field is continuous at each tissue interface. The thermal properties of materials and tissues employed are reported in **Table 3**.

The solution of Eq. (14) is a new temperature field. As previously discussed, the different system temperature determines a change in the magnetic and heat dissipation properties of the scaffolds. Also the dielectric and thermal properties of tissues vary with temperature [14]. To account for the influence of these variations on the outcome of HT treatment, the solution of Eq. (14) should be used to evaluate the EM power solving Eq. (13) for the next time step; then the next temperature distribution can be calculated considering the changed physical properties. This solution scheme is justified by the rather different dynamic of the EM and thermal fields [14].

In the temperature range 37°C–47°C of the HT treatment, the following linear relationship between the property p (ϵ, σ, C_p, k) of each tissue and the system temperature [5, 14, 25]:

$$\frac{\Delta p(T)}{p(T_0)} = 1 + c\Delta T \tag{15}$$

The dielectric properties are assumed to increase linearly with c = 3% C^{-1} [14]. The thermal properties, C_p and k, have been assumed to vary with c = 0.5%°C^{-1} and 0.33%°C^{-1}, respectively [14]. An exception was made for the heat capacity of blood, which presents a negative coefficient of −1%°C^{-1} [14].

In this condition the strength, frequency, and envelope of the external RF magnetic field required to treat both osteosarcoma and fibrosarcoma cells were investigated.

3.5 Results

The temperature pattern resulting from the exposure to the homogeneous RF field is uniform and radial, as shown in **Figure 5a**. This is a consequence of the homogeneous distribution of the MNPs in the biomaterials [7, 14]. After 60 min of treatment, it can be noticed that the temperature in the healthy bone can reach 47°C, which is a potentially harming value. To assess the performance of the two types of MagS in treating OST and FIB, the average value of temperature in the tumor region vs. time was considered. From **Figure 5b** it can be noticed that both magnetic hydroxyapatite and Fe-doped PCL are able to treat the poorly vascularized FIS for any given dimension of the residual area r_t. This can be explained considering that $P_{EM} \gg |\rho_b C_{p,b} \omega_b (T - T_a)|$, for H_0=10–17 mT. However,

Material or tissue	k, $Wm^{-1}K^{-1}$	C_p, $Jkg^{-1}K^{-1}$	Q_m, Wm^{-3}	ω_b, s^{-1}
Magnetic hydroxyapatite	1.33	700	—	—
ϵ-PCL	0.488	3359.2	—	—
Fracture gap–inflamed	0.558	2450	5262.5	$6.95 \cdot 10^{-3}$
Fracture gap–ischemic	0.558	2450	5262.5	$6.95 \cdot 10^{-3}$
Bone tumors: OST and FIB	0.32	1313	57,240	$2.42 \cdot 10^{-3} \div 0.595$
Bone	0.32	1313	286.2	$0.262 \cdot 10^{-3}$

Table 3.
Heat transfer properties of scaffolds and tissues [14].

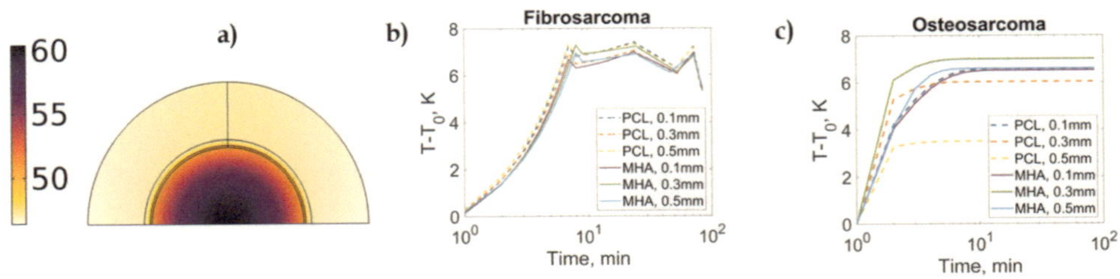

Figure 5.
(a) 2D temperature distribution after 60 min of treatment using a RF magnetic field of 30 mT and working at 293 kHz. A OST with r_t*=0.5 mm is considered. (b) Average temperature in the region with residual FIB cells. (c) Average temperature in the region with residual OST cells. (MHA = magnetic hydroxyapatite).*

the temperature rise is noticeable, and the external field should be modulated (or turned off) to keep the temperature closest to the target value of 42°. In the case of OST, give the high value of blood perfusion (see **Table 3**); not all MagS are able to treat successfully the residual cancer cells. This is the most challenging tumor. Indeed, the Fe-doped PCL scaffold fails to reach the lethal HT temperature for an OST of 0.5 mm, even increasing the amplitude to 40 mT or the frequency to 409 kHz. The magnetic hydroxyapatite scaffold is more effective in treating the residual osteosarcoma cells, as can be observed in **Figure 5c**. These results demonstrate that the in HT treatment of residual bone tumor cells is feasible, and, with the knowledge of the physico-chemical properties of the nanomaterial, the treatment can be planned against different type of tumors.

4. Magnetic scaffolds and regenerative medicine

4.1 Magnetic drug delivery

Magnetic scaffolds were conceived as a multifunctional platform for tissue engineering applications (see **Figure 1**) [1–5]. As presented in the Introduction, they are a platform for magnetically targeted drug delivery of growth factors to control and enhance tissue healing, such as in the case of bone tissue [1, 11]. The bio-nanotechnology research developed magnetic carriers of biomolecules such as VEGF or TGF-β [11, 27]; however, the problem of maximizing and controlling their delivery to the site of injury is still addressed in the literature [1–4, 13]. This section will cover the use of MagS, implanted in a damaged bone, as an in situ magnet, i.e., as attraction site for external MNPs carrying GFs. The influence on the cellular response is assessed employing a multiphysics model [13]. The prediction of the magnetic force required to attract the MNPs, the velocity in the extracellular matrix (ECM) medium, and the final spatial distribution is fundamental to foresee a treatment planning procedure, while evaluating the influence parameters in the drug delivery and the cellular migration process.

4.2 Challenging the mathematical modeling of MDD

4.2.1 The magnetostatic problem

Considering the geometry of **Figure 4**, the analysis domain is limited to the scaffold and the fracture gap, neglecting the bone tumor and assuming that only healthy bone is present, in a way similar to [13]. The MagS and the gap have a radius of 5 mm. An external uniform and static magnetic flux density field of strength B_0 is supposed to be applied along the z-axis of the system. The magnetic composite nanomaterial will magnetize in a nonlinear way according to the following relationship [13]:

$$\overline{M} = M_s\phi\left(coth\ (\zeta) - \frac{1}{\zeta}\right) \tag{16}$$

where all symbols have the previous definition. As presented in **Table 1**, the magnetization response of the scaffolds varies from a minimum of 0.4 emu·g^{-1} to a maximum of 25 emu·g^{-1}. Considering this nonlinear material property, the problem is the determination of the spatial distribution of the magnetic field, i.e., the solution of the following magnetostatic problem employing the scalar magnetic potential ψ_m [13]:

$$\begin{aligned} \nabla \times \overline{H} = 0 \\ \overline{H} = -\nabla\psi_m \end{aligned} \tag{17}$$

Due to the presence of the magnetic material, the magnetic field flux lines concentrate in the prosthetic implant, implying that the norm of the gradient of magnetic density field between the MagS and the diamagnetic tissues is relevant [6]. In the literature, it is reported that if the magnetic density field gradients are higher than 1.3 Tm^{-1}, then the magnetic force exerted on a population of surrounding MNPs would be sufficient to overcome their weight force and set them in motion toward the scaffold [13, 28]. This is a very simplified view of the problem. Indeed, several relevant physical and biological factors took part to the transport phenomena of MNP attraction to the MagS in the presence of a static magnetic field. As defined by Grief and Richardson, the magnetic force vector $\overline{F}_m$ on an ensemble of MNPs in saturation regime can be evaluated as follows [12]:

$$\overline{F}_m = \frac{M_{s2}V_{m2}}{6k_BT}\nabla\left|\overline{B}\right|^2 \tag{18}$$

where M_{s2} and V_{m2} are the saturation magnetization, in Am^{-1}, and the volume of the spherical magnetic nanoparticles, in nm^{-3}, to be attracted, respectively. The nanoparticles conjugated with growth factors or drugs are hence set in motion with a velocity v_m equal to [12, 13]:

$$\overline{v}_m = \frac{\overline{F}_m}{6\pi\eta r_{m2}} \tag{19}$$

where r_{m2} is the radius of the magnetic carriers. The term η is the viscosity of the medium in which the nanoparticles move, in Pa·s. This medium is often assumed to be water (η_w=1·10^{-3}Pa·s); however, actually the MNPs that move from the capillaries of bone tissues into the fracture gap are dragged in a solution of water, proteins (e.g., collagen, fibrin, and plasmin), and other macromolecules. Therefore, the extracellular matrix (ECM) can be assumed to be the medium in which the MNPs move, implying that η_w=1·10^{3}Pa·s [13].

After having solved Eq. (17) and calculated Eqs. (18) and (19), the spatiotemporal distribution of the concentration of MNPs (C_{mnp}, mol·m^3) functionalized with the drug can be obtained computing the following diffusion-convection equation [13]:

$$\frac{\partial C_{mnp}}{\partial t} = \nabla \cdot \left[D_{mnp} \nabla C_{mnp}\right] - \overline{v}_m \nabla C_{mnp} \tag{20}$$

D_{mnp} is the diffusion coefficient of MNPs in the medium, assumed to be equal to $10^{-9} ms^{-2}$. The analytical mass balance is subject to the outflow condition at the scaffold surface, while a constant initial concentration of MNPs ($C_{m,0}$) is assumed at the host bone interface. In the fracture gap, it is assumed that $C_{mnp}(r, z, t = 0) = 0$ as initial condition.

The magnetic field distribution (Eq. (17)) is derived by solving numerically the magnetostatic problem for the geometry depicted in **Figure 4** using the *Magnetic Fields No Currents* package from the *AC/DC module* of COMSOL Multiphysics. Then $\overline{v}_m$ is inserted in the *Transport of Dilute Species* interface to solve Eq. (20).

4.2.2 Including the cells and the biological elements

Now, we assume that the MDD system is constituted by an active GF with concentration C_{chemo} present at the MNP surface. The presence of such specific biomolecule is supposed to represent a direct chemotactic stimulus for the human mesenchymal stem cells (MSCs) present in the surrounding host bone tissue [13]. Chemotaxis is the phenomenon of cellular migration directed toward a chemical stimulus. This means that the osteoprogenitor cells can respond to the chemical signaling stimulus carried by the MNPs. The magnitude of the chemotactic stimulus can be assumed to be equal to [13]:

$$K_{chemo} = \frac{D_c}{C_{chemo}} \tag{21}$$

Given C_{mnp} from Eq. (20), the spatial pattern of the MSCs exposed to the biological signal carried by the MNPs is the solution of the mass balance for the cell population C_c:

$$\frac{\partial C_c}{\partial t} = \nabla \cdot \left[D_c \nabla C_c - C_c K_{chemo} \nabla C_{mnp}\right] \tag{22}$$

Similar to Eq. (20), Eq. (22) is subject to Dirichlet and Neumann boundary conditions, i.e., the diffusive flux of cell population should be null at the scaffold surface, and the cell concentration at host bone is set to a constant value of $C_{c,0}$. Moreover, the cell concentration in the fracture gap is assumed to be null at the initial time.

With this set of equations, it is possible to model the role of magnetic scaffolds as part of a MDD system studying the influence on the cellular migration and the scaffold colonization, providing valuable insight into the use of MagS as a tool in tissue engineering.

4.2.3 Results from the case study

The magnetic scaffolds exposed to the static magnetic flux density field B_0 respond in a way similar to a uniformly magnetized sphere, as shown in **Figure 6a**. As a matter of fact, the magnetic force and resulting velocity distribution are similar, implying that the MNP concentration is maximum at the poles of the scaffold, while it is minimum at the equator (**Figure 6b**). The generic GFs attached to the MNPs are sensed by the cells, which migrate toward the chemical stimulus. As shown by **Figure 6**, after only 24 h of migration, the MSCs invade the gap cavity

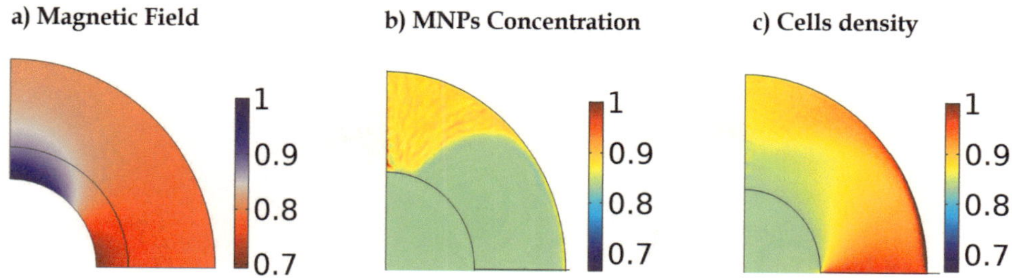

Figure 6.
(a) Normalized magnetic field distribution ($\overline{H}/H_0$). (b) Normalized MNP concentration profile after 48 h ($C_{mnp}/C_{m,0}$). (c) MSC density after 24 h ($C_c/C_{c,0}$).

and reach the biomaterial surface. This suggests that the cells can colonize the scaffold and boost the regenerative process. From the model and the results presented, the final cell density pattern can be predicted and controlled by tuning or acting on the scaffolds magnetic properties or geometry [13].

5. Conclusions

This chapter presented an innovative family of nanocomposite magnetic biomaterials and their biomedical applications. Mixing magnetic nanoparticles with traditional biomaterials, e.g., polymer or ceramics, or chemically doping them allows the manufacturing of a magnetic-responsive biomaterial with multifunctional properties. The so-called magnetic scaffolds have been studied for their ability to transduce an external magnetic signal into mechanical and biological outcome, thus proving to be a powerful platform for cell and tissue stimulation [1–4]. Exploiting the ability of the MNPs embedded in the biomaterial to dissipate power when exposed to a radio-frequency magnetic field makes MagS a valid candidate to perform local hyperthermia treatment on residual cancer cells. In this chapter the physical properties and the magnetic susceptibility of these novel composite nanosystems are investigated. Then an in silico model to study the feasibility of employing MagS in the treatment of bone cancers, such as osteosarcomas and fibrosarcomas, is presented [14]. The results indicate that further research on the nanomaterial is required to develop an effective and tailored magnetic scaffold. Finally, the potential of MagS to serve as an in vivo attraction site to enhance the magnetic drug delivery of growth factors is faced. To predict the final concentration pattern, a mathematical model which relates the nonlinear magnetic problem and the mass transport issue is presented. Furthermore, the link between these two aspects and the biological influence on cellular migration is challenged [13]. The results indicate that MagS are able to attract MNPs and exert an indirect action on MSCs in a way dependent on the geometrical and material properties.

Acknowledgements

The authors would like to sincerely thank Prof. G. Mazzarella for the helpful discussions and suggestions to this work.

Conflict of interest

The authors declare no conflict of interest.

Abbreviations

BMP-2	bone morphogenetic protein-2
DLS	dynamic light scattering
ECM	extracellular matrix
FIB	fibrosarcoma
FEM	finite element method
FF	ferrofluid
GF	growth factor
MagS	magnetic scaffold
MDD	magnetic drug delivery
MF	magnetic field
MHA	magnetic hydroxyapatite
MSC	mesenchymal stem cell
MNP	magnetic nanoparticle
OST	osteosarcoma
PCL	poly-caprolactone
RF	radio frequency
SMF	static magnetic field
TCP	tricalcium phosphate
TEM	transmission electron microscope
VEGF	vascular endothelial growth factor

Author details

Matteo Bruno Lodi* and Alessandro Fanti
Department of Electrical and Electronic Engineering, University of Cagliari, Cagliari, Italy

*Address all correspondence to: matteob.lodi@unica.it

References

[1] Liu XL et al. Magnetic nanomaterials for advanced regenerative medicine: The promise and challenges. Advanced Materials. 2018;**1804922**:1-13. DOI: 10.1002/adma.201804922

[2] Adedoyin AA, Ekenseair AK. Biomedical applications of magneto-responsive scaffolds. Nano Research. 2018;**11**(10):5049-5064. DOI: 10.1007/s12274-018-2198-2

[3] Smolkov B et al. A critical review on selected external physical cues and modulation of cell behavior: Magnetic nanoparticles, non-thermal plasma and lasers. Journal of Functional Biomaterials. 2019;**10**(1):2. DOI: 10.3390/jfb10010002

[4] Xia Y et al. Magnetic field and nano-scaffolds with stem cells to enhance bone regeneration. Biomaterials. 2018; **183**:151-170. DOI: 10.1016/j.biomaterials.2018.08.040

[5] Lodi MB. Synthesis and characterization of ferrofluid impregnated magnetic scaffolds for hyperthermia [thesis]. Politecnico di Torino; 2018. Available from: https://webthesis.biblio.polito.it/8919/1/tesi.pdf

[6] Bock N et al. A novel route in bone tissue engineering: Magnetic biomimetic scaffolds. Acta Biomaterialia. 2010;**6**(3): 786-796. DOI: 10.1016/j.actbio.2009.09.017

[7] Banobre-Lopez M et al. Hyperthermia induced in magnetic scaffolds for bone tissue engineering. IEEE Transactions on Magnetics. 2014; **50**(11):1-7. DOI: 10.1109/TMAG.2014.2327245

[8] Tampieri A et al. Intrinsic magnetism and hyperthermia in bioactive Fe-doped hydroxyapatite. Acta Biomaterialia. 2012;**8**(2):843-851. DOI: 10.1016/j.actbio.2011.09.032

[9] Iannotti V et al. Fe-doping-induced magnetism in nano-hydroxyapatites. Inorganic Chemistry. 2017;**56**(8): 4446-4458. DOI: 10.1021/acs.inorgchem.6b03143

[10] Okamoto M. The role of scaffolds in tissue engineering. In: Handbook of Tissue Engineering Scaffolds. 1st ed. Vol. 1. Sawston: Woodhead Publishing; 2019. pp. 23-49. DOI: 10.1016/B978-0-08-102563-5.00002-2

[11] Lanier OL, Monsalve AG, McFetridge PS, Dobson J. Magnetically triggered release of biologics. International Materials Reviews. 2019; **64**(2):63-90. DOI: 10.1080/09506608.2018.1446280

[12] Grief AD, Richardson G. Mathematical modelling of magnetically targeted drug delivery. Journal of Magnetism and Magnetic Materials. 2005;**293**(1):455-463. DOI: 10.1016/j.jmmm.2005.02.040

[13] Fanti A, Lodi MB, Mazzarella G. Enhancement of cell migration rate toward a superparamagnetic scaffold using LF magnetic fields. IEEE Transactions on Magnetics. 2016;**52**: 10-18. DOI: 10.1109/TMAG.2016.2583405

[14] Fanti A, Lodi MB, Vacca G, Mazzarella G. Numerical investigation of bone tumor hyperthermia treatment using magnetic scaffolds. IEEE Journal of Electromagnetics, RF and Microwaves in Medicine and Biology. 2018;**2**(4):294-301. DOI: 10.1109/JERM.2018.2866345

[15] Perez RA et al. Therapeutically relevant aspects in bone repair and regeneration. Materials Today. 2015;

18(10):573-589. DOI: 10.1016/j.mattod.2015.06.011

[16] Rosensweig RE. Heating magnetic fluid with alternating magnetic field. Journal of Magnetism and Magnetic Materials. 2002;**252**:370-374. DOI: 10.1016/S0304-8853(02)00706-0

[17] Ng EYK, Kumar SD. Physical mechanism and modeling of heat generation and transfer in magnetic fluid hyperthermia through Néelian and Brownian relaxation: A review. Biomedical Engineering Online. 2017;**16**(36):1-22. DOI: 10.1186/s12938-017-0327-x

[18] Anghileri LJ, Robert J, editors. Hyperthermia in Cancer Treatment. 1st ed. Vol. 2. Boca Raton: CRC Press; 2019. 288 p. DOI: 10.1201/9780429266546

[19] Datta NR et al. Local hyperthermia combined with radiotherapy and−/or chemotherapy: Recent advances and promises for the future. Cancer Treatment Reviews. 2015;**41**(9):742-753. DOI: 10.1016/j.ctrv.2015.05.009

[20] Landi GT, Arantes FR, Cornejo DR, Bakuzis AF, Andreu I, Natividad E. Ac susceptibility as a tool to probe the dipolar interaction in magnetic nanoparticles. Journal of Magnetism and Magnetic Materials. 2017;**421**:138-151. DOI: 10.1016/j.jmmm.2016.08.011

[21] Hergt R et al. Enhancement of ac-losses of magnetic nanoparticles for heating applications. Journal of Magnetism and Magnetic Materials. 2004;**280**(2):358-368. DOI: 10.1016/j.jmmm.2004.03.034

[22] van Berkum S, Dee JT, Philipse AP, Erne BH. Frequency-dependent magnetic susceptibility of magnetite and cobalt ferrite nanoparticles embedded in paa hydrogel. International Journal of Molecular Sciences. 2013;**14**(5):10162-10177. DOI: 10.3390/ijms140510162

[23] Peabody TD, Attar S, editors. Orthopaedic Oncology. New York: Springer International Publishing; 2016. 223 p. DOI: 10.1007/978-3-319-07323-1

[24] Santini-Araujo E, Kalil RK, Bertoni F, Park Y, editors. Tumors and Tumor-like Lesions of Bone. For Surgical Pathologists, Orthopedic Surgeons and Radiologists. London: Springer-Verlag; 2015. 994 p. DOI: 10.1007/978-1-4471-6578-1

[25] Lodi MB, Vacca G, Fanti A, Luini L, Vecchi G, Mazzarella G. Numerical comparison of magnetic biomaterials for hyperthermia applications: The osteosarcoma case. In: 2019 13th European Conference on Antennas and Propagation (EuCAP). Krakow, Poland: IEEE; 2019. pp. 1-5. Available from: http://ieeexplore.ieee.org/stamp/stamp.jsp?tp=&arnumber=8739883&isnumber=8739246

[26] Bagaria HG, Johnson DT. Transient solution to the bioheat equation and optimization for magnetic fluid hyperthermia treatment. International Journal of Hyperthermia. 2005;**21**(1):57-75. DOI: 10.1080/02656730410001726956

[27] Meikle ST et al. Surface functionalization superparamagnetic nanoparticles conjugated with thermoresponsive poly (epsilon-lysine) dendrons tethered with carboxybetaine for the mild hyperthermia-controlled delivery of VEGF. Acta Biomaterialia. 2016;**40**:235-242. DOI: 10.1016/j.actbio.2016.04.043

[28] Samal SK et al. Multilayered magnetic gelatin membrane scaffolds. ACS Applied Materials & Interfaces. 2015;7(41):23098-23109. DOI: 10.1021/acsami.5b06813

Chapter 2

Synthesis and ESR Study of Transition from Ferromagnetism to Superparamagnetism in $La_{0.8}Sr_{0.2}MnO_3$ Nanomanganite

Mondher Yahya, Faouzi Hosni and Ahmed Hichem Hamzaoui

Abstract

Electron spin resonance (ESR) spectroscopy was used to determine the magnetic state transitions of nanocrystalline $La_{0.8}Sr_{0.2}MnO_3$ at room temperature, as a function of crystallite size. Ferromagnetic nanoparticles having an average crystallite size ranging from 9 to 57 nm are prepared by adopting the autocombustion method with two-step synthesis process. Significant changes of the ESR spectra parameters, such as the line shape, resonance field (Hr), g-factor, linewidth (ΔHpp), and the low-field microwave absorption (LFMA) signal, are indicative of the change in magnetic domain structures from superparamagnetism to single-domain and multi-domain ferromagnetism by increase in the crystallite size. Samples with crystallite sizes less than 24.5 nm are in a superparamagnetic state. Between 24.5 and 32 nm, they are formed by a single-domain ferromagnetic. The multi-domain state arises for higher sizes. In superparamagnetic region, the value of g-factor is practically constant suggesting that the magnetic core size is invariant with decreasing crystallite size. This contradictory observation with the core-shell model was explained by the phenomenon of phase separation that leads to the formation of a new magnetic state that we called multicore superparamagnetic state.

Keywords: magnetic nanoparticles, perovskite manganite, ESR, ferromagnetism, superparamagnetism, multicore superparamagnetic, phase separation

1. Introduction

Magnetic nanoparticles (MNPs) display physical and chemical properties different those found in their corresponding bulk materials. These properties make them attractive in widespread applications such as energy, electronics, sensor designs of all kinds, catalysts, magnetic refrigeration, optics, and in various biomedical applications [1–6].

In medicine, MNPs has attracted attention because they are detectable, remotely manipulable, stimulable by a magnetic field, and can combine both diagnosis and therapy in one dose. These multifunctional nanomaterials can be used as contrast agents for medical imaging and nano-vectors to transport therapeutic agents to their target, in local delivery of drugs, or to destroy the cancer cells by local hyperthermia [7–10]. These magnetic platforms should possess very small size and must combine high magnetic susceptibility and loss of magnetization after removal of the magnetic field [11, 12]. The optimization of the nanoparticles' size, size distribution, agglomeration, coating, and shapes along understanding the changes in magnetic properties prompted the application of magnetic nanoparticles in diverse fields.

The magnetic properties arise from the magnetic moment associated with electron spin. In ferromagnetic materials, groups of atoms band together into areas called domains, in which all the electrons have the same magnetic orientation. Fundamental changes occur in ferromagnetic materials when their physical size is reduced. The magnetic structure of the ferromagnetic material consists of several magnetic domains and thus retains an important magnetic moment in zero fields [13]. When the particle size becomes smaller, there is a limit when it becomes energetically unfavorable with the formation of several domains and the particle becomes magnetic single domain. A single-domain particle presents all the spins aligned in the same direction. The critical size value depends on the material. A single-domain particle presents all the spins aligned in the same direction. The total magnetic moment of the nanoparticles can be regarded as one giant magnetic moment, composed of all the individual magnetic moments of the atoms. These nanoparticles show a certain preference for the direction, along which their magnetization aligns to (directions of easy magnetization). As particle size continues to decrease below a critical size, the ferromagnetic material is transformed into a superparamagnetic one. In this case, magnetization can randomly flip direction under the influence of temperature, causing the residual magnetization to be null (**Figure 1**) [14]. Superparamagnetic nanoparticles are preferred in biomedical applications because they have zero magnetization at room temperature and do not agglomerate [15, 16].

Magnetic measurements such as vibrating sample magnetometry and associated methods for determining magnetization according to the applied field and Mössbauer spectroscopy have been extensively used to observe the superparamagnetic behavior [17–19]. In this study, electron spin resonance (ESR) spectroscopy was used to study magnetic properties of $La_{0.8}Sr_{0.2}MnO_3$ nanopowders. The basic concepts of ESR are based on the Zeeman effect which leads to the separation of the energy levels of the electrons under the effect of an external magnetic field. The magnetic moments of the electrons perform a precessional motion around the direction of the applied magnetic field with the angular velocity of Larmor. Classically, the resonance event occurs when a transverse alternating field is applied at the Larmor frequency. For a ferromagnetic solid, a strong coupling exists between the electrons. In general, this energy is more

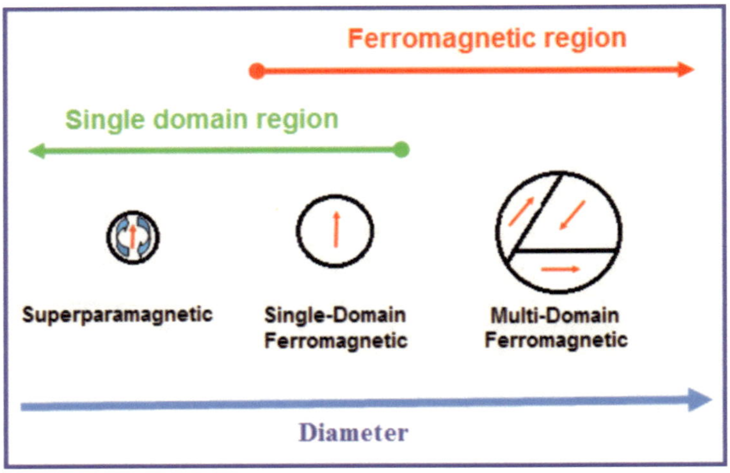

Figure 1.
Transition from multidomain to single-domain to superparamagnetic state whith increasing the particle diameter.

important than Zeeman energy. Thus, in the presence of an applied field, it is the total magnetization which will presage a precession movement [20].

ESR spectra is formed by two absorption, one to the higher fields known as the electronic magnetic resonance (EMR) and an absorption around zero magnetic field appointed low-field microwave absorption (LFMA) [21]. EMR spectrum is a resonant absorption characterized by means of two parameters, the resonant magnetic field (Hres) and the linewidth (ΔHpp). These parameters give information on magnetic nature of the materials [22]. LFMA signal is a nonresonant absorption considered as a sensitive detector of magnetic ordering [23, 24]. This signal has been used to detect the magnetic states in materials and provide highly sensitive detection of magnetic order [23]. More importantly, this signal is not present in the paramagnetic state and emerges as the temperature is lower than Curie temperature [25, 26].

Magnetic homesteads of $La_{1-x}Sr_xMnO3$ perovskites are influenced by intrinsic properties (composition and structure) and extrinsic properties (particle and crystallite sizes depending on the synthesis procedure). Their Curie temperature strongly depends on the Sr-content, and it varies between 320 and 370 K for $0.2 \le x \le 0.3$, which makes them attractive for self-controlled magnetic hyperthermia applications. Thus, by applying a high-frequency magnetic field, the magnetic nanoparticles reach their own heating temperature, which does not exceed TC [6, 27]. Their magnetization is higher than that of most materials for such applications [28].

Magnetic phase transitions as a function of temperature and crystallite size have been extensively studied in the past [29–31]. But there are controversial studies in critical sizes of the ferromagnetic- single magnetic domain-superparamagnetic transitions. The critical size is given as a function of the particle size determined by transmission electron microscopy (TEM) or by the crystallite size from DRX data calculated by the Scherrer method. For manganite particles, single magnetic domain is observed in the literature less than diameters which range between 50 and 80 nm [32–34]. Based on crystallites, the change from a multi-domain state to a single-domain state is given in the literature for sizes varying between 26 and 36 nm [34, 35], and crystallite size of the superparamagnetic transition is estimated at 11 nm [28], and between 18 and 24 nm [33, 34, 36].

The observed disparity between crystallite size and particle size is attributed to the polycrystalline nature of the particles. In this work, we will refer to the samples with their corresponding crystallite size.

For this reason, in order to determine critical sizes, we discuss crystallite size dependence of the transition from ferromagnetism to superparamagnetism at room temperature of $La_{0.8}Sr_{0.2}MnO_3$ nanoparticles by using electron spin resonance technique. These transitions are quantified by means of Hres, ΔHpp, and LFMA. To our knowledge, studies on nanoparticles of manganite with LFMA signal are scarce. Therefore, in this work, LFMA signal is used to give knowledge on magnetic state. For this purpose, to obtain the $La_{0.8}Sr_{0.2}MnO_3$ (LSrT-t) nanoparticles with different crystallite sizes, the autocombustion method was adopted with a two-step synthesis process. This method permits to prepare powders in a wide range of crystallite sizes depending on the annealing temperature (T) and heat treatment time (t).

2. Experimental methods

2.1 Autocombustion synthesis

After several years of intense research efforts, it has emerged that a large number of synthesis approaches to a wide variety of nanoparticles are available. In this work, $La_{0.8}Sr_{0.2}MnO_3$ (LSr) nanocrystalline were synthesized by the autocombustion process with two-step synthesis process. It is an interesting synthetic route for the preparation of a compound with differing crystallite sizes. The reaction is very simple and involves just lanthanum nitrate, strontium nitrate, and manganese acetate. Manganese acetate was dissolved in the minimum of distilled water under agitation. On the other hand, stoichiometric amounts of lanthanum nitrate and strontium nitrate are dissolved in distilled water under stirring. The solution was prepared by mixing aqueous solutions of ($La(NO_3)_3 + Sr(NO_3)_2$) and manganese acetate in 1:1 molar ratio. To evaporate the water, the solution was stirred in a beaker placed on a hot plate at 80°C until a viscous product was obtained. The solutions are mixed and were kept stirred in a beaker on a hot plate at 80°C to evaporate the water until you get a product with viscous appearance. After that, the gel obtained was inserted into a preheated oven at 350°C for 2 h. After a few seconds, a violent flame was produced by releasing large amounts of gas, with formation of a spongy powder of dark brown color. Finally, the powder resulting from autocombustion was calcined at 700°C under different heat treatment times and at 800, 900, and 1000°C for 15 h. The samples are designated by LSrT-t (T, calcination temperature; t, heat treatment times).

Phase analysis of all products was performed by using powder X-ray diffraction (XRD) using an X'Pert Pro PANAnalytical diffractometer with CuKα radiation (λ = 1.5418 Å) at room temperature. The crystalline phases were determined by comparison of the registered patterns with the International Center for Diffraction Data (ICDD) powder diffraction files (PDF).

The average crystallite sizes of the samples were estimated from the (0 2 4) reflections at $2\theta = 46°$, by means of the Debye–Scherrer equation: $D = K\lambda/\beta \cos\theta$, where K is a constant equal to 0.89, λ is the wavelength of the X-ray used, and β is the full width at half maxima (FWHM) of the X-ray reflection at 2θ [37]. The reflection at $2\theta = 46°$ was selected to calculate the crystallite sizes because it does not overlap with other profiles [38, 39].

Magnetization at various fields was measured at temperature 300 K using a vibrating sample magnetometer (VSM) in fields up to 5 T.

We have recorded the ESR spectra of $La_{0.8}Sr_{0.2}MnO_3$ at room temperature, with a Bruker ER-200D spectrometer, operating in the X-Band (9.30 GHz). These measurements are performed on 20 mg of loosely packed fine powder. Data were acquired with 2 mW of power, with a spectral width of 0–8000 Gauss. The ESR signal measured presents the first derivative of the microwave absorption over the magnetic field. The different magnetic state transitions are quantified by means of Landé factor (g), linewidth (ΔHpp), resonant field Hres as function of crystallite sizes, and low-field microwave absorption. The maxima and minima in the derivative signal define the peak-to-peak distance (ΔHpp); the Landé g-factor (g) is calculated based on the equation $g = h\nu/\mu B Hres$ where h is Planck's constant, ν is microwave frequency, μB is Bohr magneton, and Hres defines the resonance field determined by the zero-crossing point (dp/dH = 0) [40].

2.2 Structural properties

Figure 2(a) shows the XRD patterns of the precursor and the powders calcined at 700°C (LSr700- t). It seems that all patterns share the same characteristic peaks, indicating that the manganite with perovskite structure has been formed for all samples. Their widths confirm that the $La_{0.8}Sr_{0.2}MnO_3$ powders obtained were in the form of nanocrystals. For precursor powder (sample without any extra heat treatment LSrp), the characteristic peaks of perovskite phase appear with small extra peaks at θ = 25.6, 30 and 39° related to La2O3 phase (JCPDS 83-1355); hence we did not include it in our studies. The high heat released during autocombustion leads to forming the perovskite phase with small amounts of impurities. Calcination at 700°C during 0.5 h is sufficient to obtain the pure phase. This shows that precursor powder obtained from the autocombustion is very reactive. All peak positions were indexed with perovskite-type crystalline structure of $La_{0.8}Sr_{0.2}MnO_3$ (JCPDS 53–0058) with a rhombohedral space group R-3c. By increasing the heat treatment times, the width of the diffraction peaks decreased, suggesting an increase in the crystallite size. The evolution curve of the average sizes given by the Scherrer formula as a function of the heat treatment times is of sigmoid shape (**Figure 2(b)**). The sizes rapidly increase with treatment durations less than 5 h, to reach an

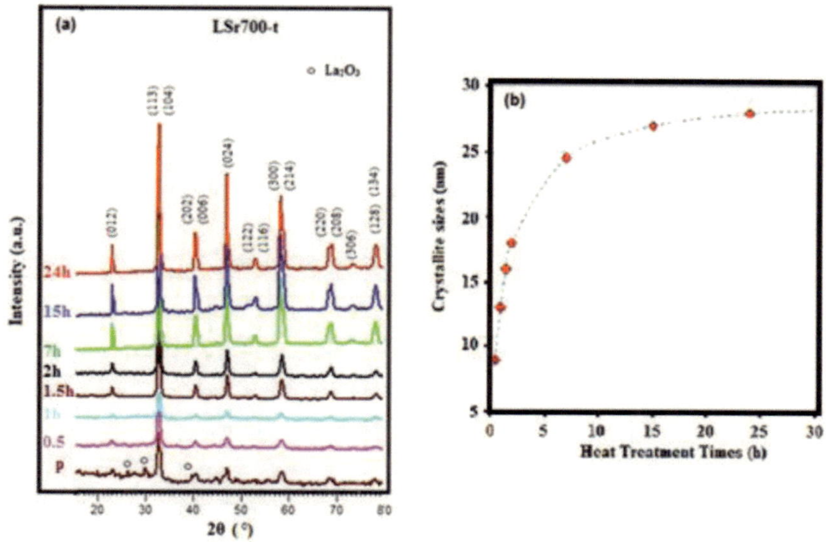

Figure 2.
(a) X-ray diffraction patterns of the samples prepared at 700°C for different processing times. (b) Variation of crystallite size with heat treatment time [20].

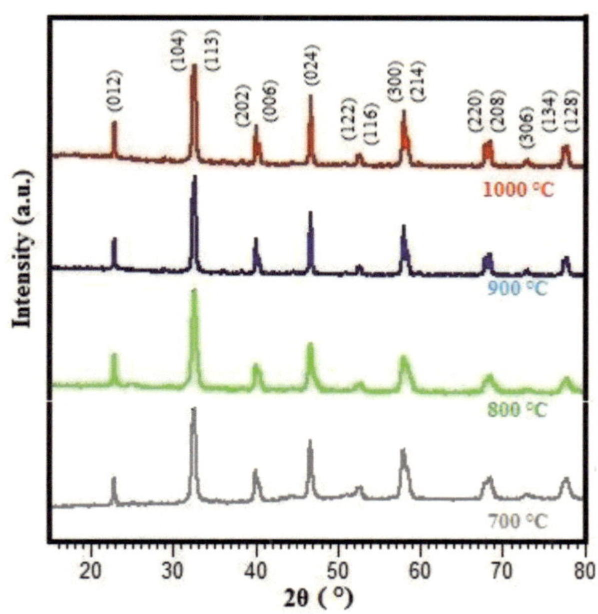

Figure 3.
X-ray diffraction patterns of samples calcined for 15 h at 700, 800, 900 and 1000°C.

asymptote at 28 nm for times greater than 15 h. By increasing the calcination temperature to 800, 900, and 1000°C for 15 h, the widths of the peaks gradually decreased (**Figure 3**). Their crystallite sizes are equal to 32, 55, and 57 nm, respectively.

The autocombustion method offers the advantage of being an exothermic process, self-propagated, and initiated at low temperature. The exothermic reaction between acetate and nitrate ions leads to the formation of the perovskite phase. The nucleation by rearrangement of short-range networks of neighboring atoms is favored by heat treatment. Modified heat treatment conditions such as the temperature and the duration of the heat treatment allowed to prepare a nanocrystalline powder of sizes between 9 and 57 nm [20].

3. Magnetic studies

The ESR spectra obtained from room temperature of all LSrT-t samples are shown together in **Figure 4**. In the first observation, the absorption signal changes radically with crystallite sizes. Two limiting cases are distinguished. For the sizes smaller than

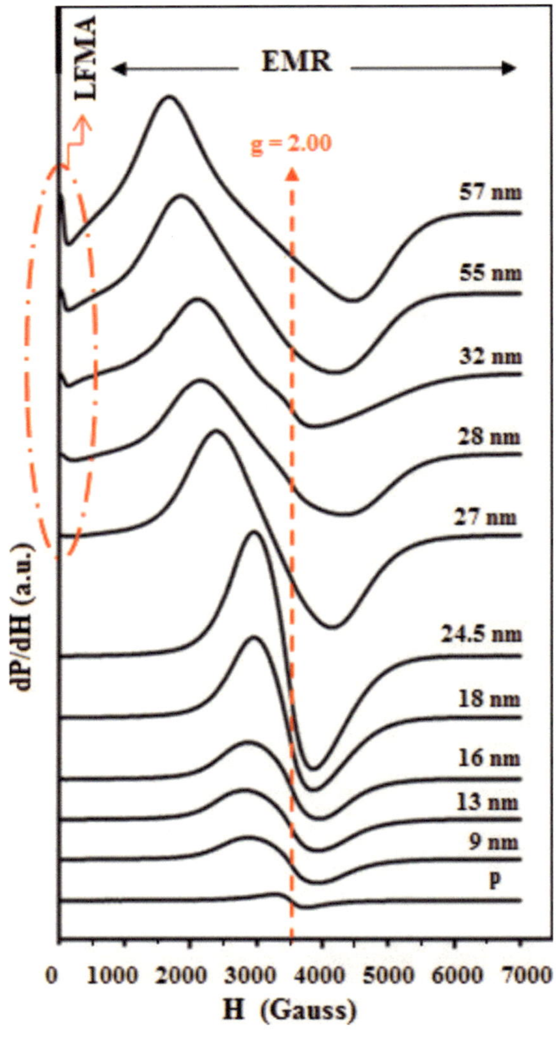

Figure 4.
ESR spectra measured at room temperature for $La_{0.8}Sr_{0.2}MnO_3$ samples with different crystallite sizes [20].

27 nm, the spectra are formed by a single narrow line observed at the higher magnetic field (greater than 1500 G), which can be straightforwardly associated with the electron magnetic resonance absorption (EMR). The EMR is specifically associated with resonant absorption and hence resonant transitions between energy levels by Zeeman effect. This absorption is symmetric with a Lorentzian line shape. Resonance field position stays roughly constant around 3500 G, with g-factor ~ 2 for samples not exceeding 24.5 nm. These spectra are comparable to those of perovskite manganite in a parametric state observed above the Curie temperature. For sizes larger than 27 nm, signal is formed by two absorptions. A strong absorption at high field corresponds to EMR absorption. An additional absorption in the low-field range less than 1000 G is associated with the low-field microwave absorption. As the crystallite size increases, the EMR absorption mode changes toward a broad asymmetric line of Dyson-type and resonance field shifts toward lower values [19, 41].

EMR is a resonant absorption that satisfies Larmor's condition defined by the expression $\omega = \gamma$ H, where ω is the resonance frequency, γ is the gyromagnetic factor, and H is the total field on the spins.

In the ferromagnetic samples with spontaneous magnetization resulting from parallel alignment of spins by magnetic interaction effect, the internal fields Hint is added to the applied field giving rise to a total field H = H0 + Hint [42]. The resonance condition is reached at low values of external field to satisfy the Larmor relation [21, 23].

Strontium doping transforms $La_{1-x}Sr_xMnO3$ manganites into a ferromagnetic state at room temperature, due to the reinforcement of the double exchange (DE) interactions between Mn^{3+} and Mn^{4+} ions, which is a ferromagnetic interaction. The ferromagnetic transition depends on the strontium content [43, 44]. The Curie temperature (TC) increases monotonically with x, and TC is increased from 220 to 325 K for values of x between 0 and 0.2 [18, 45].

Zoom of the spectral region from 3100 to 3700 G **Figure 5(a)**) shows that the samples with crystallite sizes below 28 nm have a resonance field (Hres) between 3200 and 3500 G (g-factor between 2.04 and 2.21). These values are lower than the typical value of paramagnetic manganites which is equal to Hres = 3528 G (g = 1.98) [46]. In addition to that, g-factor is bigger than the value of the free electron (ge = 2.0023) [47].

All the samples are attracted by a magnet which confirms the ferromagnetic state. Thus, the presence of the internal field leads to the displacement of the resonance field toward the weak field compared to that of the paramagnetic manganites. On the other hand, the decrease of Hres with decreasing crystallite size is an indication of a reduction in the internal field. Ferromagnetic manganite nanoparticles of few tens of nanometers are formed by a core-shell structure [34]. The core is a ferromagnetic volume enveloped by a magnetically dead layer that contains most of the oxygen defects and the faults in crystallographic structure.

When the particle size is reduced to the nanoscale, two sources contribute to decreasing ferromagnetic volume of manganite. First, the percentage of dead layer increases with increased surface/volume ratio. In addition to that, the shell thickness increases and the volume of the magnetic core decreases [34, 48]. Magnetization of the shell is considered as null, and the contributory portion of each crystallite to the magnetization is the core.

On the other hand, the state of agglomeration in the polycrystalline nanoparticles has an effect on the macroscopic magnetic properties [19]. Considering that the crystallites are in intimate contact, the increase in the shell thickness decreases the magnetic exchange energy between the two cores of neighboring particles, which promotes and improves their separation. In addition, the decrease in the size of the magnetic volume provides another source of new properties. In the nanometric state under the effect of the size reduction, the ferromagnetic material goes from a

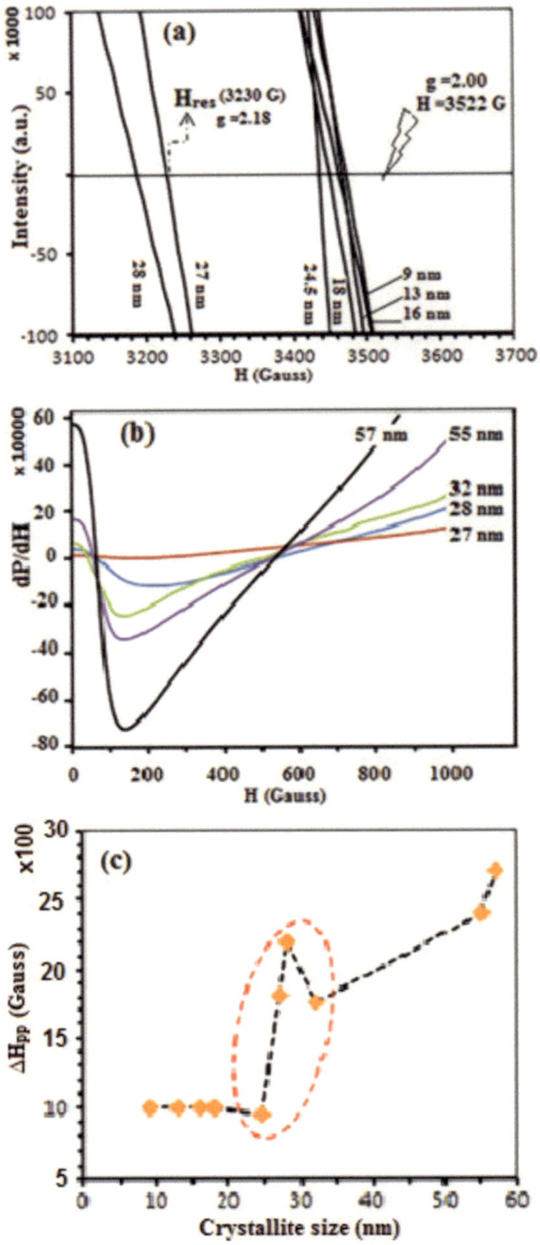

Figure 5.
Variation depending on the crystallite size of: (a) resonance fields in the spectral zone of 3100–3700 G, (b) low field microwave absorption and (c) linewidth ΔHpp [20].

multi-domain state to a magnetic single-domain state and then to the superparamagnetic state. The changes observed in the EMR and LFMA signal appear to be indicative of these magnetic state transitions.

4. Low-field microwave absorption

A zoom on the low-field spectral region (**Figure 5(b)**) shows that the intensity of LFMA gradually decreases with decreasing crystallite size and disappears from 27 nm. This microwave absorption, around zero fields, is a nonresonant absorption.

In ferromagnetic materials, magnetic domains are in a fragile state of equilibrium, and the Bloch wall which is a narrow transition region at the boundary between magnetic domains moves with very low applied fields. In fact, LFMA is associated with dynamics of the magnetic domains in material [49]. The existence of this absorption at room temperature is an indication of the ferromagnetic state of the material used to detect magnetic order. For bulk samples, the LFMA signal is used to determine the Curie temperature of ferromagnetic compounds [23, 50]. Above 28 nm the existence of LFMA shows that these compounds are in a magnetic multi-domain state and a flat response for compounds with smaller size shows that they are in a single-domain state. According to Montiel et al., the absence of LFMA signal in the ferromagnetic compounds is a good indication of superparamagnetic state in the samples [21, 22]. By size reduction, the nanocrystalline passes from a particle with several magnetic domains to a monodomain particle; the latter is either in a single-domain state or in a superparamagnetic state. Low-field absorption cannot determine the intermediate state.

Particles belonging to the single-domain state are characterized by maximum magnetocrystalline anisotropy energy; consequently the direction of magnetization is "frozen.' This characteristic has an effect on the linewidth of resonant absorption; a comparative analysis can reveal the critical size of changes in magnetic states.

5. Electronic magnetic resonance

The peak-to-peak EMR linewidth ΔHpp is an important parameter in measuring magnetic property.

ΔHpp may be due to various factors, namely, magnetic anisotropy field (ΔHK), sample porosity (ΔHpor), demagnetization field (ΔHD), and eddy currents (ΔHeddy) [51]. In the polycrystalline particles, the crystallites are randomly oriented; in that case, the contribution of the magnetocrystalline anisotropy field is dominant and we can have the following approximation Hpp=HK [52].

Linewidth ΔHpp plotted as a function of the crystallite sizes shown in **Figure 5(c)** can be subdivided into three regions. The first part, corresponding to samples with crystallite sizes less than 24.5 nm, ΔHpp has a low value around 1000 G, and between 27 and 32 nm and ΔHpp greatly increases and goes through a maximum at 28 nm. Finally, for sizes greater than 32 nm, ΔHpp increases with a lower slope. This curve calls back the variation of coercivity (Hc) with the particle size, which is maximal for particles in a single-domain state [18].

The magnetocrystalline anisotropy energy in the superparamagnetic state is small and comparable to thermal energy. By random fluctuations of the magnetization due to thermal excitation, the directions of easy magnetization vanish. This is reflected in the low value of ΔHpp [53]. A narrow resonance line is considered as the fingerprint of superparamagnetism at high temperatures, where the energy barrier is dominated by thermal oscillations [54, 55]. Thus, particles smaller than 24.5 nm are in a superparamagnetic state.

Important resonance broadening occurring at superparamagnetic zone boundary indicate that the samples with crystallites size 27, 28, and 32 nm are single-domain ferromagnetic. In the particle formed by a single magnetic domain, the magnetocrystalline anisotropy energy is proportional to the magnetic volume (EB = K V, where K and V are the anisotropy constant and volume of the particle) [56]. In the single-domain region, energy barrier separating two directions of easy magnetization is high. The magnetization requests more energy to get itself aligned along the applied field. The angular dependence of the ΔHpp results in significant increases in linewidth [42].

In the multi-domain state, application in the measurement of weak magnetic field can easily move the magnetic domain walls. The magnetization vectors approach the direction of the applied field, thus leading to reduce ΔHpp. Thus, the decrease of ΔHpp in addition to the appearance of the absorption at low field confirms multi-domain state of samples higher size 32 nm.

These results are comparable to the overall results obtained with samples of analogous compositions [35, 57]. In particular, Sujoy R. et al. reported an exponential increase of the magnetic core between 8 and 22 nm [18], favoring the transition from superparamagnetic state to the single-domain ferromagnetic state.

In addition to the intrinsic causes of line broadening due to the change of magnetic state, extrinsic causes related to the size and shape of the magnetic particles are to be considered. Nanocrystalline powders are formed by several crystallites bonded together to form a wide variety of sizes and shapes that have different magnetic properties [58, 59]. "Crystallite size" is not synonymous with "particle size"; X-ray diffraction is sensitive to the size of crystallite inside the particles.

To study the effect of particle sizes on the line shape, ESR measurements have been performed for two populations with diverse particle size selected by magnetic separation from the sample LSr900- 15 h (55 nm).

6. Effect of particle size

6.1 Magnetic separation of the nanoparticles

The force exerted on a magnetic particle placed in a static magnetic field is proportional to its volume (FM~Vp) [58]. So, according to their sizes, magnetic nanoparticles dispersed in a liquid can be efficiently separated by static magnetic field. In the experiment, the LSr900-15 h powder is dispersed in distilled water with mechanical stirring and then subjected to a magnetic field applied by a permanent magnet. The powder retained after 5 minutes is designated by LSr900b (bigger), and the powder retained from the supernatant after 30 minutes is designated by LSr900s (smaller).

6.2 Structural characterization sand ERS measurement

Figure 6 shows X-ray diffraction patterns of powders LSr900b and LSr900s with LSr900-15 h. The diffraction peaks are characteristic of the perovskite phase. The broadening of the (0 2 4) reflection at $2\theta = 46°$ is in harmony with that of the original compound LSr900-15 h. Thus, the crystallite's size for all the compounds is 55 nm.

The ESR spectra obtained for the three samples have the same general shape (**Figure** 7). They are characteristic of a multi-domain ferromagnetic compound. In coherence with the results of Valenzuela et al., the LFMA signal becomes more intense with increased particle sizes [24]. The linewidth ESR depends on particle sizes, shows a gradual decrease with decreasing size, and is explained by external causes resulting from a distribution of the anisotropy axes [60]. In fact, when a ferromagnetic particle is exposed to a magnetic field, the crystallites tend to orient in the direction of easy magnetization. The assembly of these crystallites in agglomerates prevents some of them from being oriented in the preferred direction, which translates to the spectrum level by the widening of the ESR lines. These observations confirm that the absorption line envelops ensembles of narrower and indistinguishable lines, each of them coming from the resonance of a set of spins called by Portis "spin packets" [61].

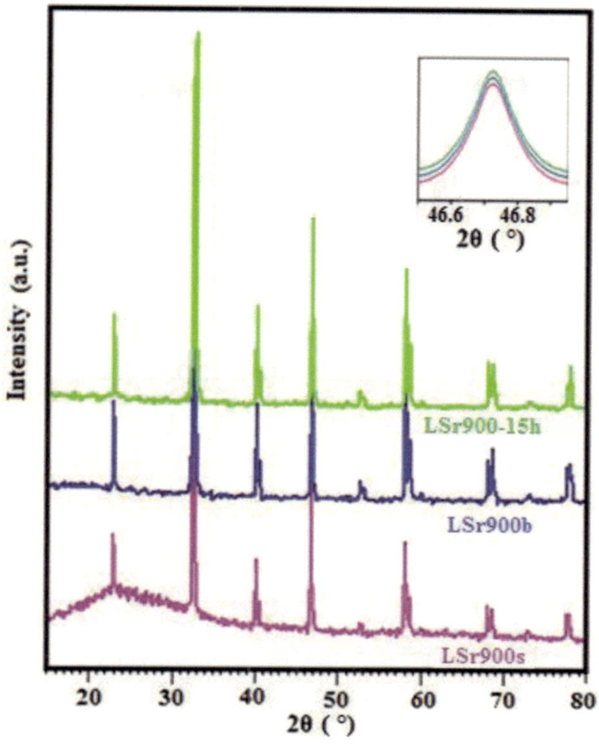

Figure 6.
X-ray diffraction patterns of the samples LSr900-15h, LSr900b and LSr900s. Inset shows the details of the (0 2 4) reflection.

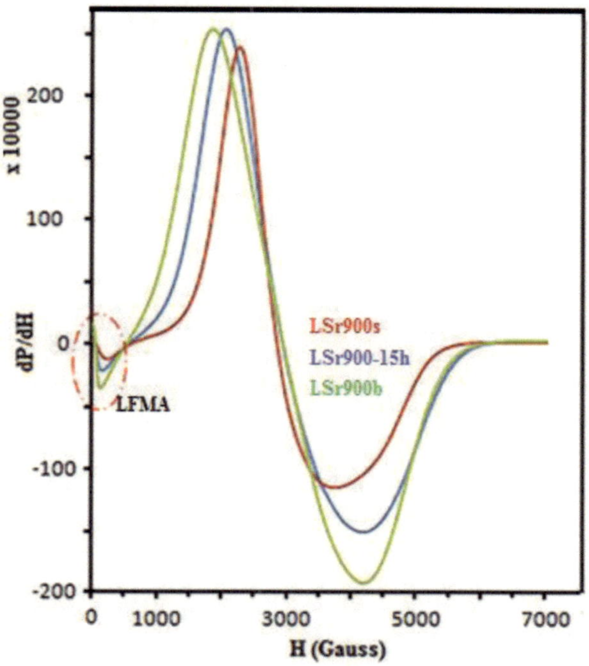

Figure 7.
ESR spectra of the samples LSr900-15h, LSr900a and LSr900b.

However, the resonance frequency remains constant, equal to Hres = 2920 G for the three samples. This can be explained in terms of the same internal field, which is directly related to the crystallite size without exchange interaction between adjacent crystallite. Autocombustion synthesis gives voluminous powders of high specific surfaces, with a structure of "sponge" form [62, 63]. These structures weaken the exchange interactions between the crystallites. In addition to this, the formation of a magnetically dead layer, which increases in thickness with decreasing crystallite size, prevents crystallite-crystallite interactions [34, 62, 64]. Thus, the resonance field is controlled principally by crystallite size.

7. Dependence of the g-factor with the crystallite size

7.1 Detection of magnetic states transitions by g-factor

The Landé g-factor is a dimensionless number calculated from the resonance field based on equation $g = h\nu/\mu BHres$. The curve of g-factor variation as a function of crystallite size can be subdivided into three parts (**Figure 8**). The first is with g values around 2.05, corresponding to the samples with crystallite sizes less than 24.5 nm. This value is slightly higher than that of the typical value of perovskite manganite materials in a paramagnetic state (g = 1.996) [46]. A strong increase in g-factor was observed in the intermediate region corresponding to the samples with crystallite sizes of 24.5 and 32 nm. In the last region, corresponding to the samples with larger crystallite sizes above 32 nm, the g-factor value continues to increase with a lower slope.

The observed slope changes coincide with those reported variations of the linewidth and other ESR spectrum parameters (line shape, LFMA, etc.). So, these changes reflect transitions of magnetic states, from superparamagnetism to ferromagnetism and from single-domain to multi-domain.

To confirm this hypothesis, magnetization measurements were recorded at 300 K and 360 K, collected in a vibrating sample magnetometer (VSM), for the three samples (a) LSr700-2 h (16 nm), (b) LSr700- 24 h (28 nm), and (c) LSr900-15 h (55 nm), belonging to different regions (**Figure 9**).

At room temperature (300 K), the nonlinearity in magnetization curves confirms the ferromagnetic behavior of three samples, while the linearity of the magnetization curve recorded at 360 K reflects a typical paramagnetic character [32]. The magnetization for samples (a) reaches more than 90% of its experimental saturation (Ms) value just above 0.5 T. Moreover, the very small residual magnetization (Mr) considered to be negligible reflects the superparamagnetic character of this sample [63, 65]. The existence of a thick dead magnetic layer in the superparamagnetic state explains the decrease of the saturation magnetization [13, 17].

A magnetization/demagnetization curve of the samples (b, c) with hysteresis loops characterizes the materials in a ferromagnetic state. For sample (b) there is no complete saturation even in the vicinity of 5 T. This means that the magnetic moments are blocked and their alignment toward the applied field requires larger values; this characterizes single-domain ferromagnetic particles [13, 17] Moreover, magnetic remanence (Mr) practically equals 0.5 Ms.; this value is consistent with the model established by Stoner and Wohlfarth for single-domain ferromagnetic particles [66, 67].

Significant reduction in Ms. for the sample (a) compared to samples (b, c) is coherent with both ESR measurements (low factor g) and core-shell model [68]. The core-shell model assumes the formation of a magnetically dead layer of thickness t, which increases with size reduction and can be estimated using the following formula [48]:

$$t = \frac{d}{2}\left(1 - \frac{M_{s\,(nano)}}{M_{s\,(bulk)}}\right)^{1/3} \tag{1}$$

where t is dead layer thickness and d is crystallite size.

As a necessary condition for superparamagnetism, nanoparticles must be formed by small magnetic volumes without exchange interaction [68]. The previous formula shows a significant increase in the dead layer for sizes below 24.5 nm, which has favored superparamagnetism. ESR spectra for these compounds have similarities: narrow linewidth, symmetrical form, g-factor less than 2.08, and absence of LFMA.

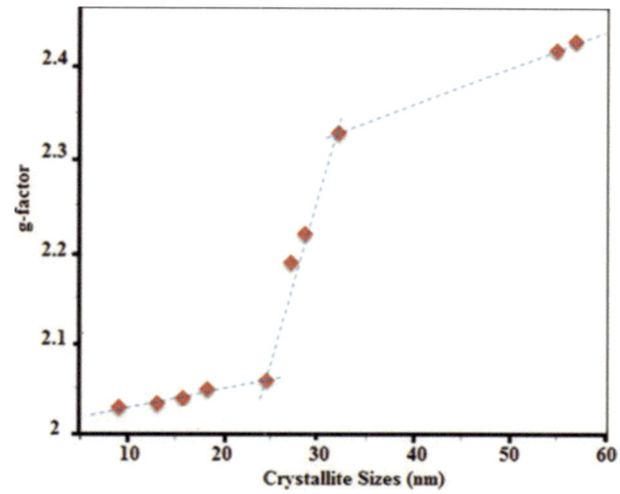

Figure 8.
The change of g-factor with crystallite size.

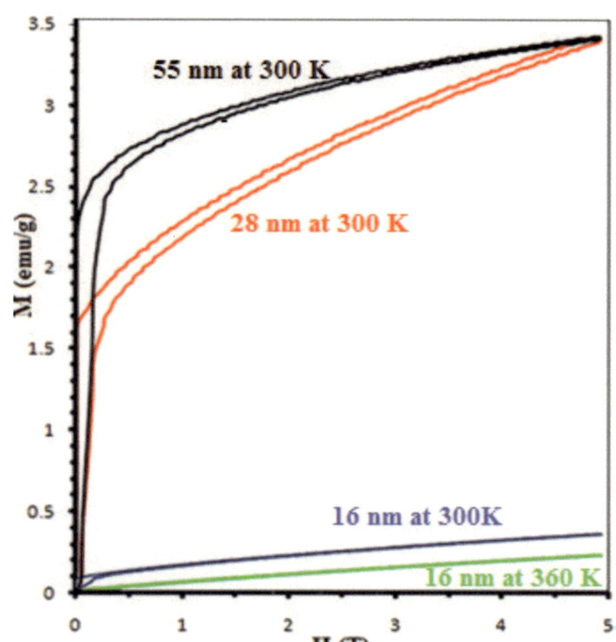

Figure 9.
Magnetization/demagnetization curves of the samples with the crystallite sizes 28 and 55 nm recorded at 300 K. and for sample of 16 nm recorded at 300 K and 360K.

The increase in the magnetic volume due to a sharp decrease in the thickness of the dead layer between 24.5 and 32 nm breaks the superparamagnetism. The system evolves toward the single-domain state, which results in a significant increase of the g-factor and ΔHpp. Beyond 32 nm, nanoparticles change into a multi-domain state. This passage is confirmed by the asymmetrical EMR signal and the appearance of the LFMA.

These results are similar to those of the literature for manganite with the perovskite structure of similar composition [18, 35, 58].

7.2 Superparamagnetic crystallite with multicore

In **Figure 8**, we notice that in the powders with crystallite size less than 24.5 nm (superparamagnetic region), the value of the factor g is practically constant. This suggests that the magnetic core size is invariant regardless of crystallite size.

To rely on the core-shell model with a single magnetic core, in order that the magnetic core keeps the same volume, the thickness of the dead layer must increase with increasing crystallite size. This hypothesis is in contradiction with the literature which reveals that the thickness of the shell decreases [18, 34, 48]. To explain these experimental data, we resorted to the phenomena of magnetic phase separation, which is a usual phenomenon observed in manganites [69, 70]. The competition between several interactions in perovskite manganites means that there are only small energy differences between the different possible phases of the system. As a result, the perovskite manganite oxide is magnetically inhomogeneous, consisting of different spatial regions with different magnetic orders.

Phase separation leads to the formation of a nanometer-sized ferromagnetic droplet cloud ranging from tens of nanometers to several hundred nanometers [71, 72]. In particular, $La_{1-x}Sr_xMnO3$ manganites tend to form mixed phases near the paramagnetic-ferromagnetic phase transition [65, 73]. In this case, transition from the single-domain to the superparamagnetic state is not due to increasing the thickness of the dead layer but to the subdivision of the magnetic volume in small volumes. Thus, the superparamagnetic particles are formed by a paramagnetic volume containing several distinct magnetic cores that we have called superparagmanetic multicore crystallite.

The multicore superparamagnetic state has a comparable magnetic structure with multicore magnetic particles (MCP)that have been shown to be promising for a wide range of biomedical applications, in particular in magnetic hyperthermia treatment and magnetic resonance imaging [74, 75]. So, we can conclude that control of crystallite size in the synthesis of manganites allows a controlled change of magnetism in these compounds.

8. Conclusion

In recent years, a remarkable progress has been made in understanding the magnetic phenomenon in nanomanganites with perovskite structure. This advance is mainly made possible through experimental measurements and theoretical approaches. In this work, ESR spectroscopic measurements were adopted to study crystallite size-dependent magnetic properties of $La_{0.8}Sr_{0.2}MnO_3$ nanomanganite. Samples with different crystallite sizes ranging from 9 to 57 nm were prepared by autocombustion method with a two-step synthesis process. Significant differences in the ESR spectrum parameters, namely, resonant field (Hres), line shape, low-field microwave absorption, and linewidth (ΔHpp), are used to determine the critical sizes of magnetic state changes. The findings from the ESR measurements are confirmed by VSM measurements. Anomaly observed in the superparamagnetic state has been explained by the formation of multicore magnetic crystallites. These results show the potential use of nanoparticles with crystallite size less than 24.5 nm for biomedical applications. Multicore superparamagnetic state has interesting magnetic properties and can have a large heating capacity equivalent to the magnetic nanoflowers.

Author details

Mondher Yahya[1*], Faouzi Hosni[2] and Ahmed Hichem Hamzaoui[1]

1 Laboratory of Useful Materials Valuation, National Center for Research in Materials Sciences, Soliman, Tunisia

2 Faculty of Sciences, University of Bisha, Bisha, Saudi Arabia

*Address all correspondence to: mondher.yahya@cnrsm.rnrt.tn

References

[1] Hussain BHKI, Anwar MS, Kim JW, Chung KC. Influence of La addition on the structural, magnetic and magnetocaloric properties in Sr2−xLaxFeMoO6 (0x0.3) double perovskite. Ceramics International. 2016;**42**:13098-13103

[2] Lee HS, Park CS, Park HH. Effect of La3+ substitution with Gd3+ on the resistive switching properties of La0.7Sr0.3MnO3 thin films. Applied Physics Letters. 2014;**104**:0-5. DOI: 10.1063/1.4876115

[3] Xu D, Luo L, Ding Y, Xu P. Sensitive electrochemical detection of glucose based on electrospun La0.88Sr0.12MnO3 naonofibers modified electrode. Analytical Biochemistry. 2015;**489**:38-43. DOI: 10.1016/j.ab.2015.08.013

[4] Nandwana V, Ryoo SR, Kanthala S, McMahon KM, Rink JS, Li Y, et al. High-density lipoprotein-like magnetic nanostructures (HDL-MNS): Theranostic agents for cardiovascular disease. Chemistry of Materials. 2017;**29**:2276-2282. DOI: 10.1021/acs.chemmater.6b05357

[5] Mornet S, Vasseur S, Grasset F, Veverka P, Goglio G, Demourgues A, et al. Magnetic nanoparticle design for medical applications. Progress in Solid State Chemistry. 2006;**34**:237-247. DOI: 10.1016/j.progsolidstchem.2005.11.010

[6] Pollert E, Veverka P, Veverka M, Kaman O, Závěta K, Vasseur S, et al. Search of new core materials for magnetic fluid hyperthermia: Preliminary chemical and physical issues. Progress in Solid State Chemistry. 2009;**37**:1-14. DOI: 10.1016/j.progsolidstchem.2009.02.001

[7] Berkova Z, Jirak D, Zacharovova K, Lukes I, Kotkova Z, Kotek J, et al. Gadolinium- and manganite-based contrast agents with fluorescent probes for both magnetic resonance and fluorescence imaging of pancreatic islets: A comparative study. ChemMedChem. 2013;**8**:614-621. DOI: 10.1002/cmdc.201200439

[8] Eguchi A, Mulya A, Lazic M, Radhakrishnan D, Berk MP, Povero D, et al. Microparticles release by adipocytes act as "find-me" signals to promote macrophage migration. PLoS One. 2015;**10**:1-19. DOI: 10.1371/journal.pone.0123110

[9] Sato I, Umemura M, Mitsudo K, Fukumura H, Kim JH, Hoshino Y, et al. Simultaneous hyperthermia-chemotherapy with controlled drug delivery using single-drug nanoparticles. Scientific Reports. 2016;**6**:1-12. DOI: 10.1038/srep24629

[10] Pillai G, Cox A, Yuen L. The science and technology of cancer theranostic nanomedicines: A primer for clinicians and pharmacists. SOJ Pharmacy & Pharmaceutical Sciences. 2018;**5**:1-17. DOI: 10.15226/2374-6866/5/2/00178

[11] Rosensweig RE. Heating magnetic fluid with alternating magnetic field. Journal of Magnetism and Magnetic Materials. 2002;**252**:370-374

[12] Vines JB, Yoon J-H, Ryu N-E, Lim D-J, Park H. Gold nanoparticles for photothermal cancer therapy. Frontiers in Chemistry. 2019;**7**:1-16. DOI: 10.3389/fchem.2019.00167

[13] Leslie-Pelecky DL, Rieke RD. Magnetic properties of nanostructured materials. Chemistry of Materials. 1996;**8**:1770-1783. DOI: 10.1021/cm960077f

[14] Busquets M, Espargaró A, Sabaté R, Estelrich J. Magnetic nanoparticles cross the blood-brain barrier: When physics rises to a challenge. Nanomaterials.

2015;**5**:2231-2248. DOI: 10.3390/nano5042231

[15] Lee JS, Cha JM, Yoon HY, Lee JK, Kim YK. Magnetic multi-granule nanoclusters: A model system that exhibits universal size effect of magnetic coercivity. Scientific Reports. 2015;**5**:1-7. DOI: 10.1038/srep12135

[16] Sharma VK, Waldner F. Superparamagnetic and ferrimagnetic resonance of ultrafine Fe3O4particles in ferrofluids. Journal of Applied Physics. 1977;**48**:4298-4302. DOI: 10.1063/1.323418

[17] Chérif W, Ellouze M, Lehlooh A-F, Mahmood SH, Elhalouani F. Structure, magnetic properties and Mössbauer spectra of La0.67Sr0.33F xMn1-xO3 manganites oxide prepared by mechanical ball milling method. Hyperfine Interactions. 2012;**211**:153-164

[18] Sujoy R, Dubenko I, Edorh DD, Ali N. Size induced variations in structural and magnetic properties of double exchange La0.8Sr0.2MnO3-δ nano-ferromagnet. Journal of Applied Physics. 2004;**96**:1202-1208. DOI: 10.1063/1.1760230

[19] Hintze CE, Fuchs D, Merz M, Amari H, Kübel C, Huang M, et al. Size-induced changes of structural and ferromagnetic properties in La1-x SrxMnO3 nanoparticles. Journal of Applied Physics. 2017;**121**:214303. DOI: 10.1063/1.4984829

[20] Yahya M, Hosni F, M'Nif A, Hamzaoui AH. ESR studies of transition from ferromagnetism to superparamagnetism in nano-ferromagnet La0.8Sr0.2MnO3. Journal of Magnetism and Magnetic Materials. 2018;**466**:341-350. DOI: 10.1016/j.jmmm.2018.04.050

[21] Montiel H, Alvarez G, Conde-gallardo A, Zamorano R. Microwave absorption behavior in Cr2O3 nanopowders. Journal of Alloys and Compounds. 2015;**628**:272-276. DOI: 10.1016/j.jallcom.2014.11.198

[22] Montiel H, Alvarez G, Conde-Gallardo A, Zamorano R. Effect of the particle size on the microwave absorption in the yttrium-iron garnet. Journal of Nano Research. 2014;**28**:73-81. DOI: 10.4028/www.scientific.net/JNanoR.28.73

[23] Alvarez G, Montiel H, Barron JF, Gutierrez MP, Zamorano R. Yafet-Kittel-type magnetic ordering in Ni0.35Zn0.65Fe2O4 ferrite detected by magnetosensitive microwave absorption measurements. Journal of Magnetism and Magnetic Materials. 2010;**322**:348-352. DOI: 10.1016/j.jmmm.2009.09.056

[24] Valenzuela R, Herbst F, Ammar S. Ferromagnetic resonance in Ni-Zn ferrite nanoparticles in different aggregation states. Journal of Magnetism and Magnetic Materials. 2012;**324**:3398-3401. DOI: 10.1016/j.jmmm.2012.02.051

[25] Owens FJ. DC magnetic field dependent microwave absorption in CMR material, La0.7Sr0.3MnO3. Journal of Physics and Chemistry of Solids. 1997;**58**:1311-1314. DOI: 10.1016/S0022-3697(97)00040-1

[26] Owens FJ. Ferromagnetic resonance observation of a phase transition in magnetic field- aligned Fe2O3 nanoparticles. Journal of Magnetism and Magnetic Materials. 2009;**321**:2386-2391. DOI: 10.1016/j.jmmm.2009.02.135

[27] Apostolov A, Apostolova I, Wesselinowa J. Specific absorption rate in Zn-doted ferrites for self-controlled magnetic hyperthermia. European Physical Journal B. 2019;**92**:58

[28] Rashid AU, Ahmed A, Ahmad SN, Shaheen SA, Manzoor S. Study of specific absorption rate of strontium

doped lanthanum manganite nanoparticles for self-controlled hyperthermia applications. Journal of Magnetism and Magnetic Materials. 2013;**347**:39-44. DOI: 10.1016/j.jmmm.2013.07.045

[29] Owens FJ. Ferromagnetic resonance observation of a phase transition in magnetic field- aligned Fe2O3nanoparticles. Journal of Magnetism and Magnetic Materials. 2009;**321**:2386-2391. DOI: 10.1016/j.jmmm.2009.02.135

[30] Channu VSR, Holze R, Walker EH. Nanostructures for Solid Oxide Fuel Cells. New Journal of Glass and Ceramics. 2013;**2013**:29-33. DOI: 10.4236/njgc.2013.31005

[31] Autret-Lambert C, Jirak Z, Gervais M, Poirot N, Gervais F, Raimboux N, et al. Electron spin resonance and neutron diffraction studies of Nd. Physica B: Condensed Matter. 2007;**19**:5222-5229

[32] Curiale J, Sánchez RD, Troiani HE, Leyva AG, Levy P. Room-temperature ferromagnetism in La2/3Sr13MnO3 nanoparticle assembled nanotubes. Applied Physics Letters. 2005;**87**:43113. DOI: 10.1063/1.1999842

[33] Vázquez-vázquez C, Blanco MC, López-quintela MA, Sánchez RD, Rivas J, Oseroff SB. Characterization of La0.67Ca0.33MnO3±δ particles prepared by the sol–gel route. Journal of Materials Chemistry. 1998;**8**:991-1000

[34] Dey P, Nath TK. Effect of grain size modulation on the magneto- and electronic-transport properties of La0.7Ca0.3Mn O3 nanoparticles: The role of spin-polarized tunneling at the enhanced grain surface. Physical Review B: Condensed Matter and Materials Physics. 2006;**73**:1-14. DOI: 10.1103/PhysRevB.73.214425

[35] Dutta A, Gayathri N, Ranganathan R.Effect of particle size on the magnetic and transport properties of La0.875Sr0.125MnO3. Physical Review B. 2003;**68**:54432. DOI: 10.1103/PhysRevB.68.054432

[36] Pollert E, Kaman O, Veverka P, Veverka M, Marysko M, Záveta K, et al. Core-shell La(1-x)Sr(x)MnO3 nanoparticles as colloidal mediators for magnetic fluid hyperthermia. Philosophical Transactions. Series A, Mathematical, Physical, and Engineering Sciences. 2010;**368**:4389-4405. DOI: 10.1098/rsta.2010.0123

[37] Wilson JILAJC. Seherrer after sixty years: A survey and some new results in the determination of crystallite size. Journal of Applied Crystallography. 1978;**11**:102-113

[38] Tang YK, Sui Y, Xu DP, Qian ZN, Su WH. Study of magnetoresistance in nanostructured La2/3Sr 1/3MnO3 powder compacts. Journal of Magnetism and Magnetic Materials. 2006;**299**:260-264. DOI: 10.1016/j.jmmm.2005.04.009

[39] Turky AO, Rashad MM, Hassan AM, Elnaggar EM, Bechelany M. Optical, electrical and magnetic properties of lanthanum strontium manganite La1-xSrx MnO3 synthesized through the citrate combustion method. Physical Chemistry Chemical Physics. 2017;**19**:6878-6886. DOI: 10.1039/c6cp07333f

[40] Phan TL, Thanh TD, Phan MH, Yu SC. Electron spin resonance spectra of La0.7Cd0.3(Mn, Co)O3 Perovskites. IEEE Transactions on Magnetics. 2014;**50**:1-4. DOI:10.1109/TMAG.2013.2273084

[41] Dyson FJ. Electron spin resonance absorption in metals. II. Theory of electron diffusion and the skin effect. Physical Review. 1958;**98**:349-359

[42] Datt G, Kotabage C, Abhyankar AC. Ferromagnetic resonance of NiCoFe2O4 nanoparticles and microwave absorption properties of flexible NiCoFe 2 O 4 –carbon black/poly(vinyl alcohol) composites. Physical Chemistry Chemical Physics. 2017;**19**:20699-20712. DOI: 10.1039/C7CP03953K

[43] Kinoshita T, Furuyabu T, Adachi M. Synthesis of La 0.75Sr0.25MnO3 fine particles for self- controlled magnetic heating hyperthermia by ultrasonic spray pyrolysis. Japanese Journal of Applied Physics. 2014;**53**:70302. DOI: 10.7567/JJAP.53.070302

[44] Moradi J, Ghazi ME, Ehsani MH, Kameli P. Structural and magnetic characterization of La0.8Sr0.2MnO3 nanoparticles prepared via a facile microwave-assisted method. Journal of Solid State Chemistry. 2014;**215**:1-7. DOI: 10.1016/j.jssc.2014.03.011

[45] Khanduri H, Dimri MC, Vasala S, Leinberg S. Magnetic and structural studies of LaMnO 3 thin films prepared by atomic. Journal of Physics D: Applied Physics. 2013;**46**:175003. DOI: 10.1088/0022-3727/46/17/175003

[46] Causa MT, Alejandro G, Zysler R, Prado F, Caneiro A, Tovar M. Jahn-teller effects on the superexchange interactions in LaMnO3. Journal of Magnetism and Magnetic Materials. 1999;**196**:506-508. DOI: 10.1016/S0304-8853(98)00856-7

[47] Taiwo FA. Electron paramagnetic resonance spectroscopic studies of iron and copper proteins. Spectroscopy. 2003;**17**:53-63. DOI: 10.1155/2003/673567

[48] Moradi J, Ghazi ME, Ehsani MH, Kameli P. Structural and magnetic characterization of La0.8Sr0.2MnO3nanoparticles prepared via a facile microwave-assisted method. Journal of Solid State Chemistry. 2014;**215**:1-7. DOI: 10.1016/j.jssc.2014.03.011

[49] Valenzuela R, Alvarez G, Montiel H, Gutiérrez MP, Mata-Zamora ME, Barrón F, et al. Characterization of magnetic materials by low-field microwave absorption techniques. Journal of Magnetism and Magnetic Materials. 2008;**320**:1961-1965. DOI: 10.1016/j.jmmm.2008.02.008

[50] Gavi H, Ngom BD, Beye AC, Strydom AM, Srinivasu VV, Chaker M, et al. Low- field microwave absorption in pulse laser deposited FeSi thin film. Journal of Magnetism and Magnetic Materials. 2012;**324**:1172-1176. DOI: 10.1016/j.jmmm.2011.11.003

[51] Pacewicz A, Krupka J, Salski B, Aleshkevych P, Kopyt P. Rigorous broadband study of the intrinsic ferromagnetic linewidth of monocrystalline garnet spheres. Scientific Reports. 2019;**9**:9434. DOI: 10.1038/s41598-019-45699-7

[52] Srivastava CM, Patni MJ. Ferromagnetic relaxation processes in polycrystalline magnetic insulators. Journal of Magnetic Resonance (1969). 1974;**15**:359-366. DOI: 10.1016/0022-2364(74)90090-0

[53] Byun TY, Byeon SC, Hong KS, Kim CK. Origin of line broadening in Co-substituted NiZnCu ferrites. Journal of Applied Physics. 2000;**87**:6220-6222

[54] Ivanshin VA, Deisenhofer J, Krug von Nidda H-A, Loidl A, Mukhin AA, Balbashov AM, et al. ESR study in lightly doped La1-xSrxMnO3. Physical Review B. 2000;**61**:6213-6219. DOI: 10.1103/PhysRevB.61.6213

[55] Desai P, Song K, Koza J, Pariti A, Nath M. Soft-chemical synthetic route to superparamagnetic FeAs@C core-shell nanoparticles exhibiting high blocking temperature. Chemistry of Materials. 2013;**25**:1510-1518. DOI: 10.1021/cm303632c

[56] Chandra S, Das R, Kalappattil V, Eggers T, Harnagea C, Nechache R, et al. Epitaxial magnetite nanorods with enhanced room temperature magnetic anisotropy. Nanoscale. 2017;**9**: 7858-7867. DOI: 10.1039/c7nr01541k

[57] Krivoruchko VVV, Konstantinova T, Mazur A, Prokhorov A. Magnetic resonances spectroscopy of nanosize La0.7Sr0.3MnO3. Journal of Magnetism and Magnetic Materials. 2006;**300**:e122-e125

[58] Kowalczyk B, Lagzi I, Grzybowski BA. Strategies for size and/or shape-selective purification of nanoparticles. Current Opinion in Colloid & Interface Science. 2011;**16**:135-148. DOI: 10.1016/j.cocis.2011.01.004

[59] Reddy LH, Arias JL, Nicolas J, Couvreur P. Magnetic nanoparticles: Design and characterization, toxicity and biocompatibility, pharmaceutical and biomedical applications. Chemical Reviews. 2012;**112**:5818-5878

[60] Usselman DJSRJ, Russek SE, Klem MT, Allen MA, Douglas T, Young M, et al. Temperature dependence of electron magnetic resonance spectra of iron oxide nanoparticles mineralized in listeria innocua protein cages. Journal of Applied Physics. 2012;**112**:84701

[61] Bowman MK, Hase H, Kevan L. Saturation behavior of inhomogeneously broadened EPR lines detected with magnetic field modulation. Journal of Magnetic Resonance. 1976;**22**:23-32

[62] Yang JYYJ, Wen TL, Tu H, Wang DQ. Characteristics of lanthanum strontium chromite prepared by glycine nitrate process. Solid State Ionics. 2000;**135**:475-479

[63] Bulte JAFJWM, Brooks RA, Moskowitz BM, Bryant LH. Relaxometry and magnetometry of the MR contrast agent MION-46L. Magnetic Resonance in Medicine. 1999;**42**(2):379-384

[64] Zhu T, Shen BG, Sun JR, Zhao HW, Zhan WS. Surface spin-glass behavior in La2/3Sr1/3MnO3 nanoparticles. Applied Physics Letters. 2001;**78**:3863-3865. DOI: 10.1063/1.1379597

[65] Zhang XGLT, Wang XP, Fang QF. Magnetic and charge ordering in nanosized manganites magnetic and charge ordering in nanosized manganites. Applied Physics Reviews. 2014;**1**:31302

[66] Morales AMRMA, Finotelli PV, Coaquira JAH, Rocha-Leão MHM, Diaz-Aguila C, Baggio-Saitovitch EM. In situ synthesis and magnetic studies of iron oxide nanoparticles in calcium-alginate matrix for biomedical applications. Materials Science and Engineering: C. 2008;**28**:253-257

[67] Schrefl JFT, Hrkac G, Suess D, Scholz W. Coercivity and remanence in self-assembled FePt nanoparticle arrays. Journal of Applied Physics. 2003;**93**:7041-7043

[68] Markovich V, Fita I, Wisniewski A, Jung G, Mogilyansky D, Puzniak R, et al. Spin-glass-like properties of La0.8Ca0.2 MnO3 nanoparticles ensembles. Physical Review B: Condensed Matter and Materials Physics. 2010;**81**:134440. DOI: 10.1103/PhysRevB.81.134440

[69] Rivadulla JRF, López-Quintela MA. Origin of the glassy magnetic behavior of the phase segregated state of the perovskites. Physical Review Letters. 2004;**93**:167206

[70] Dagotto AME, Hotta T. Colossal magnetoresistant materials: The key role of phase separation. Physics Reports. 2001;**344**:1-153

[71] Liang L, Li L, Wu H, Zhu X. Research progress on electronic phase

separation in low- dimensional perovskite manganite nanostructures. Nanoscale Research Letters. 2014;**9**:325. DOI: 10.1186/1556-276X-9-325

[72] Saurel CSD, Brûlet A, Heinemann A, Martin C, Mercone S. Magnetic field dependence of the magnetic phase separation in Pr1-xCaxMnO3 manganites studied by small-angle neutron scattering. Physical Review B. 2006;**73**:94438

[73] Ulyanov TEKAN, Yang DS, Mazur AS, Krivoruchko VN, Levchenko GG, Danilenko IA. Local structure and magnetic inhomogeneity of nano-sized La 0.7Sr0.3MnO3 manganites. Journal of Applied Physics. 2011;**109**:0-4

[74] Wells J, Bogart LK, Pankhurst Q, Wells J, Johansson C, Posth O, et al. Standardisation of magnetic nanoparticles in liquid suspension. Journal of Physics D: Applied Physics. 2017;**50**:383003. DOI: 10.1088/1361-6463/aa7fa5

[75] Bender LNP, Bogart LK, Posth O, Szczerba W, Rogers SE, Castro A. Structural and magnetic properties of multi-core nanoparticles analysed using a generalised numerical inversion method. Scientific Reports. 2017;7:45990

Chapter 3

Chiral Hybrid Nanosystems and Their Biosensing Applications

Vladimir E. Bochenkov and Tatyana I. Shabatina

Abstract

The presented chapter is devoted to chiral biosensing using various metal nanostructures and their hybrid nanosystems with optically active bio- and organic molecules. Plasmonic nanosystems and nanostructures provide an excellent platform for label-free detection of molecular adsorption by detecting tiny changes in the local refractive index or amplification of light-induced processes in biomolecules. Based on recent theoretical and experimental developments in plasmon-enhanced local electric fields, we consider the main types of molecular-plasmonic hybrid systems capable of generating an amplified chiroptical signal for such applications as detecting the presence of certain biomolecules and (in some cases) determination of their orientation and higher-order structure.

Keywords: plasmonics, chiral nanosystems, biosensors, surface plasmon resonance, chirality

1. Introduction

Plasmonic nanostructures attract increasing attention as signal amplifiers and transducers for optical sensing. The local plasmon-induced enhancement of electric fields affects various optical processes in molecular systems and therefore finds multiple applications in enhanced spectroscopic techniques, such as Surface-Enhanced Raman Scattering (SERS), Plasmon-Enhanced Fluorescence (PEF), Surface-Enhanced Infrared Absorption (SEIRA), etc. These plasmon-enhanced spectroscopic methods have been described in several recent review articles [1–3].

A wide number of naturally occurring biomolecules, including amino acids, sugars, and nucleotides, are chiral and often exist only in one of the two possible enantiomeric forms [4–6]. It is known that the therapeutic effect of chiral drugs is associated with one of enantiomer's form, while the other one can lead to highly undesirable effects [6]. This fact highlights the necessity of enantiomers detection and separation as well as the need for stereoselective synthesis. Circular dichroism (CD) spectroscopy is usually used for analyzing molecular chirality. It determines small differences in the interactions between left- and right-circularly polarized light (CPL) with a molecular system. Molecular chirality can also reveal itself by rotating a polarization plane of linearly polarized light, an effect called optical rotational dispersion (ORD). These chiroptical responses define each other through Kramers-Kronig relations [7]. Unfortunately, their experimental detection is possible only in highly concentrated analytes because the chiroptical signals are usually very weak. Thus, there is great interest in chiral plasmonics nowadays, especially in using plasmon-enhanced near fields in order to increase the sensitivity of chiroptical spectroscopy [8, 9]. Many chiral plasmonic nanostructures are studied both experimentally and theoretically. The nature of plasmonic chirality and the chiroptical effects in plasmonic nanostructures are described in several extensive reviews [10–12].

This chapter is dedicated to the applications of different plasmonic metal nanostructures and their hybrid nanosystems with optically active organic and biomolecules in chiral biosensing. We focus on the recently published results of using plasmon-induced evanescent fields and consider the main types of molecular plasmonic systems capable of generating amplified chiroptic signal in order to detect the presence of certain biomolecules and (in some cases) to determine their orientation and high-order structure.

2. Chiral molecules coupled to a single plasmonic nanoantenna

The absorbance of an achiral excitonic system can be strongly enhanced via the interaction with plasmonic particles due to the antenna effect [13]. Similar phenomena can amplify the chiroptical response of a chiral molecular system coupled to a plasmon, which is highly desirable for (bio)sensing, as it can increase the sensitivity and lower the detection limit of a sensor.

It was experimentally demonstrated that silver nanoparticles induce nearly two orders of magnitude amplification of chromophore's CD. Besides, the CD signal appeared at the same wavelength as the Localized Surface Plasmon Resonance (LSPR) of the nanoparticles [14]. These results initiated thorough theoretical studies of the nature of plasmon-induced CD amplification [15–18]. It has been found that binding of a chiral molecule to a small achiral plasmonic nanoparticle (particle size ≈ 10 nm) can enhance the molecular CD signal in conjunction with the appearance of a new CD signal at the plasmon resonance wavelength. The suggested model is presented in **Figure 1**.

The charge is localized in the small gap between the molecule and the nanoparticle and the Coulomb interactions between their electronic systems can be very strong. As a result, the CD signal induced by molecular transition can be boosted by plasmonic

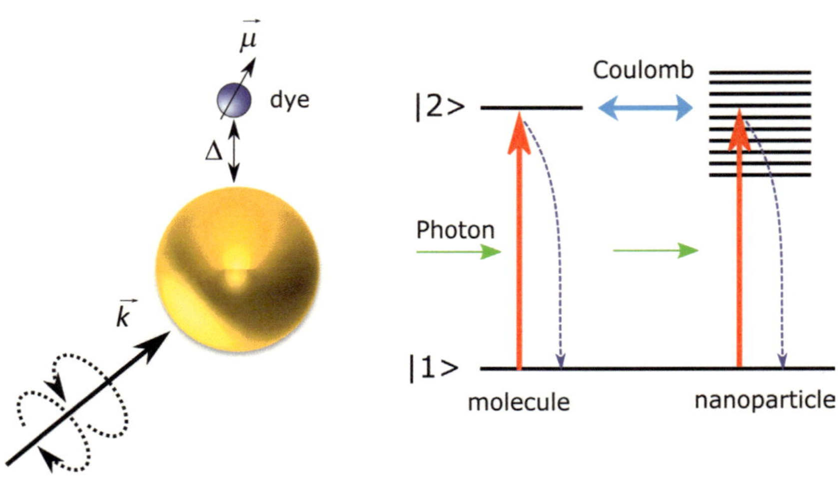

Figure 1.
The model of the metal nanoparticle-dye molecule assembly (left) and the scheme of excitation-deexcitation processes (right) with solid vertical (horizontal) arrows representing light (Coulomb)-induced transitions. The dotted vertical arrows show the relaxation processes. Adapted with permission from Govorov et al. [15], Copyright 2010 American Chemical Society.

resonance, and, at the same time, the surface current in plasmonic nanoparticle becomes chiral due to the chiral molecule, resulting in a new CD signal [15]. With increasing distance (d) between the metal nanoparticle and the molecule, the near field intensity decreases as d^{-3}. Since most biomolecules have CDs in the UV range, the appearance of the CD peak in the visible range is important for practical purposes as it can be easily detected.

A number of experimental studies support these findings. The CD peak in the visible range, generated by an adsorption of peptide molecules on spherical gold nanoparticles with 10 nm radius, is presented in **Figure 2** [19].

Similar results were obtained for a bilayer of riboflavin 50-monophosphate and polylysine molecules adsorbed on gold island films [20]. For cysteine and its derivatives in the presence of 45 nm spherical silver nanoparticles [21], for glutathione molecules adsorbed on 45 nm silver nanocubes [22] and for tobacco mosaic virus uniformly covered by spherical 5 nm gold nanoparticles [23] the enhanced chiroptical signals in the visible spectral range were also detected. The authors of the latter work achieved near-field amplification of the induced CD signal controlling by the separation between the molecule and the metal nanoparticle. The adsorbed chiral molecules were found to be in resonance with plasmon for the new appeared CD peak. Though, matching the wavelengths of molecular absorption and plasmon resonance intensified the effect, which suggests the use of nanostructures with plasmon resonance in UV for biosensing [24, 25].

Thus, to amplify CD signal of small chiral molecules a plasmonic nanoantenna can be used. But practical application of this approach in biosensing is limited by strong sensitivity to separation distance. It was shown that the induced CD can be sensitive to the orientation of the transition dipole moments of the adsorbed molecule relative to the local plasmon polarization of the nanoparticle [26]. Two orders of magnitude enhancement coefficients of CD were observed for 42 ± 2 nm gold/silver core-shell nanocubes coated with DNA molecules [24]. Additionally, the induced CD signal can be sensitive to the orientation of the transition dipole moments of the adsorbed molecule relatively to the local plasmonic polarization of the nanoparticle [26].

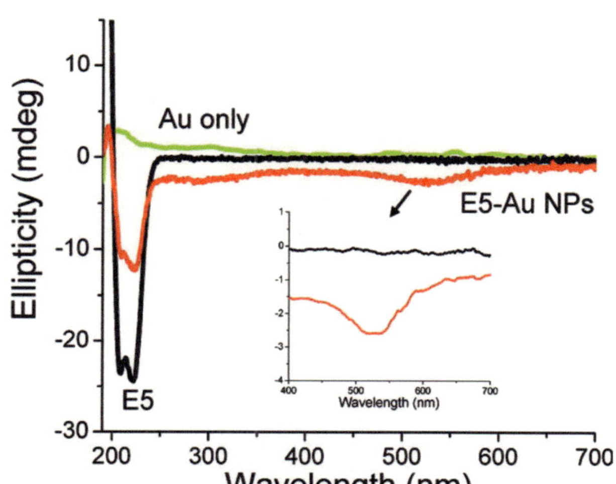

Figure 2.
CD spectra of E5 peptide (black line), gold nanoparticles (green line), and their assembly (red line). The inset shows the appearance of the CD peak in the visible range. Reproduced with permission from Slocik et al. [19]. Copyright 2011 American Chemical Society.

3. Chiral molecules in the gap of dimers of plasmonic metal nanoparticles

A similar chirooptical phenomenon is theoretically predicted for chiral molecules located in the gap of dimers of plasmonic nanoparticles for both resonant and nonresonant CD signals [27–30]. Strongly amplified electric field localized in the gap of such nanostructures results in a strong CD amplification. The enhancement factors up to 3000 can be expected for 5 nm gap. **Figure 3a,b** demonstrate the schematic image of the system and the correlation of CD enhancement with electric field in the gap [31].

This effect was experimentally observed in several systems of aggregated metal nanoparticles (NP), such as spherical gold nanoparticles conjugated to oligonucleotides [32–34], cysteine-modified gold nanorods [35, 36] and nanospheres [37, 38], as well as cholate-coated silver nanoparticle [39], etc. The CD signal is greatly enhanced by particle agglomeration, which leads to the formation of nanometer-sized gaps.

A demonstration of chiral biosensing based on the formation of asymmetric plasmon dimers has been reported recently [40]. The authors detected CD signal generated by self-assembly of gold nanorods using an antigen-antibody recognition reaction. The low detection limit for bisphenol A of 0.02 ng/mL was reached. To assemble nanoplasmon dimers with a tunable chiroptical response, the immobilization of highly specific complementary DNA primers on their surface were

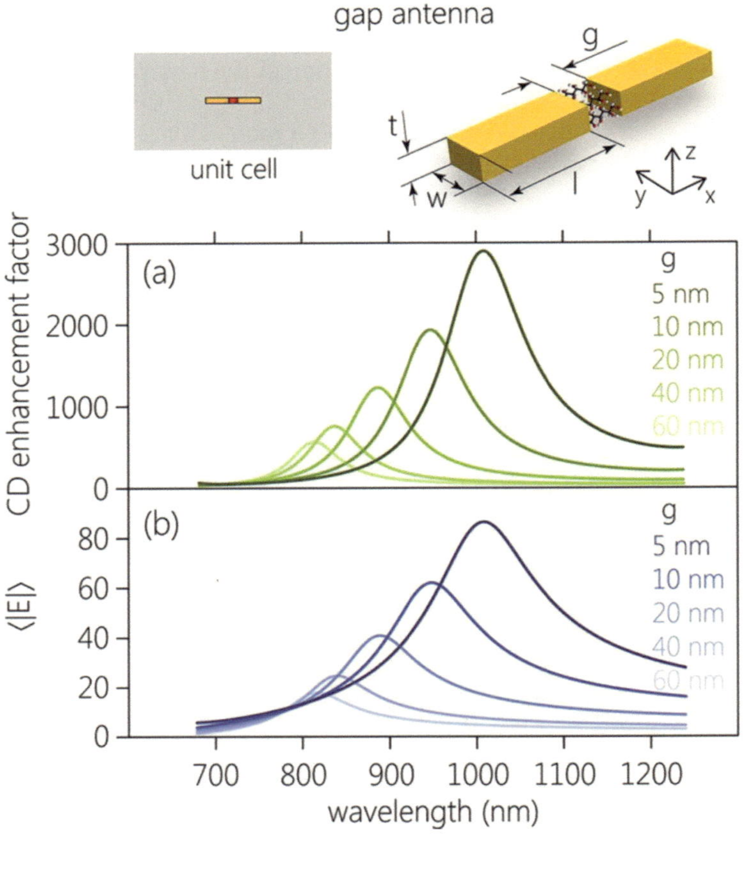

Figure 3.
Chiral molecules in the gap of a plasmonic dimer: schematics (top), (a) calculated CD enhancement factor for different gap size; (b) corresponding values of average electric field in the gap volume. Reproduced with permission from Nesterov et al. [31].

fulfilled [34]. Using this method and its modifications, various analytes were successfully being detected. For instance, self-assembly of silver ions was carried out with a detection limit of 2 pM by inducing the nanodimer assembly reaction [41]. By triggering the DNA-based chiroplasmonic assembly of gold nanoparticles and endonuclease H-paII the methyltransferase activity and inhibition has been studied [42]. Self-assembly technique has been used for the detection of ochratoxin A with the reported limit of detection as low as 0.15 pg/mL. In this experiment the construction of heterogeneous system with Au core and Ag satellites was created [43]. For the demonstration of the zeptamolar DNA detection the multimetal shell-engineered assemblies had been used [44].

Recently an alternative sensing scheme was demonstrated on an example of the alpha-fetoprotein (AFP) detection. NP dimers were formed by AFP aptamer binding to its complementary sequence immobilized on gold nanoparticles and display strong chiroptical activity. The addition of AFP aptamer led to its strong binding to immobilized aptamers and to the destruction of the aptamer-DNA hybrid accompanied by the decrease of CD signal. The authors reported the detection limit as low as 11 pg mL^{-1}. Since AFP is a marker of liver cancer, such detection scheme is of high practical interest [45].

Similar approach was used for the detection of 8-hydroxy-2′-deoxyguanosine, a well-known biomarker for the oxidative DNA damage, in human serum [46]. CD intensity showed log-linear correlation with the concentration of the analyte molecules, and the limit of the detection as 33 pM was reported.

The same strategy has been applied for monitoring the intracellular concentration of analyte molecules. For instance, the detection of adenosine-5′-triphosphate

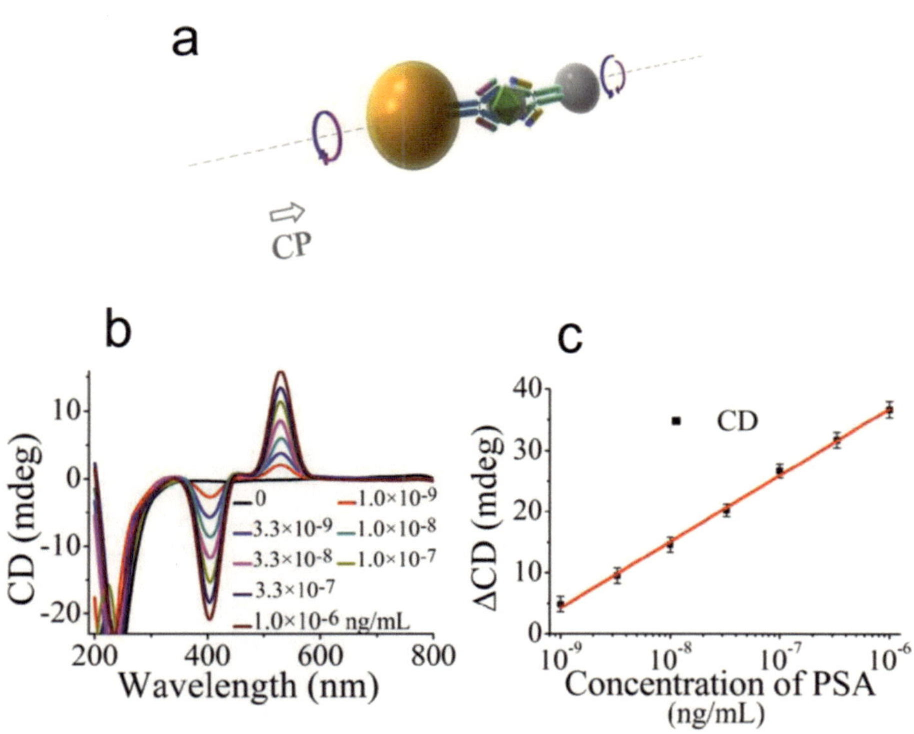

Figure 4.
Schematics of the Au-Ag heterodimer bridged by immunocomplex with prostate-specific antigen (PSA) (a); (b) the CD spectra for different concentrations of PSA; (c) the CD calibration curve for PSA detection as a function of logarithmic PSA concentrations. Reproduced with permission from Wu et al. [32], Copyright 2013 American Chemical Society.

in living cells with a detection limit of 0.2 mM has been demonstrated [47]. The results of telomerase activity study were reported: a limit of detection as 1.7×10^{-15} IU in a single HeLa cell was reached [48]. **Figure 4a–c** demonstrates another general approach to the formation of sensitive plasmonic dimers by using antibody-antigen interactions. For the chiroplasmonic detection of an environmental toxin, microcystin-LR, and a cancer biomarker, prostate-specific antigen (PSA), a silver-gold nanoparticle heterodimer was used [32]. The PSA limit of detection as 5×10^{-10} ng/mL (1.5×10^{-20} M) has been determined.

Recently, the detection of DNA molecules (with concentrations below 100 pM) located in the gap of two gold nanoparticles was successfully demonstrated [49]. The effects of the particle's gap size, shape, etc. were analyzed. In this way, for the detection of small chiral molecules, enzymes and proteins plasmonic dimers can be used. This approach is highly promising for biomedical applications such as biorecognition and intracellular detection. The use of DNA contributes to high sensitivity and selectivity in practical applications.

4. Chiral assemblies of achiral plasmonic nanoparticles

Consider sensing approach basing on geometrical chirality of chiral molecules. Physically, the generation of chiroptical response is due to both exciton-plasmon interaction, as described in previous section, or plasmon-plasmon coupling between achiral nanoparticles, arranged in a chiral superstructure [50]. In the latter case, the molecules play a role of a template and define handedness of the resulting assemblies. Eventually, the presence of the chiral molecules can be detected by the appearance of the CD signal.

Currently, several techniques were used to arrange plasmonic metal nanoparticles in chiral superstructures. As an example, DNA molecules can serve as such a template and chiral DNA-based structures are considered in recent reviews [51–56]. This method is often called "DNA origami." Besides, promising results have been achieved using peptide-directed assembly of gold nanoparticles into single-helix [57, 58] and double-helix superstructures [59, 60] (see **Figure 5**).

In order to arrange metallic nanoparticles in chiral assemblies, cholesteric liquid crystals were also used by the authors [61–63]. The elongated nanofibers obtained in cholesteric liquid crystalline mesophase are demonstrated in **Figure 6** [62]. Firstly, silver nanoparticles (2.5 nm in size) were formed and then their helical ordering was resulted due to the specific interactions of ligand molecule's functional thiol- groups with the surface atoms of silver nanoparticles. This process was resulted in hybrid spirals formation. In the case of a thiocholesterol ligand, concentrated organosols of such aggregates have high optical activity, as demonstrated by CD spectra. The optical spectra of these systems exhibited the absorption at about 430–450 nm, attributed to the LSPR bands of silver nanoparticles, and long-wavelength absorption at 700–1000 nm, which is characteristic of linear aggregates. This near infrared region is suitable for in vivo imaging and biosensing.

Anisotropic plasmonic nanoparticles, such as nanorods, offer some additional advantages, including better near-field focusing and high sensitivity to relative particle orientation [60]. The generation of optical activity in a dimer of gold nanorods was theoretically investigated. It was shown that hybrid plasmon modes, binding and antibinding, contribute to the CD signal [64]. Increased chirooptic activity was shown also by dimers and larger aggregates of gold nanorods initiated by immobilized DNA oligomers during PCR [65], which made it possible to detect DNA attomols [66]. The chiral response arose due to a slight twist between the nanorods. In the process of self-assembly, both enantiomers were formed, but one

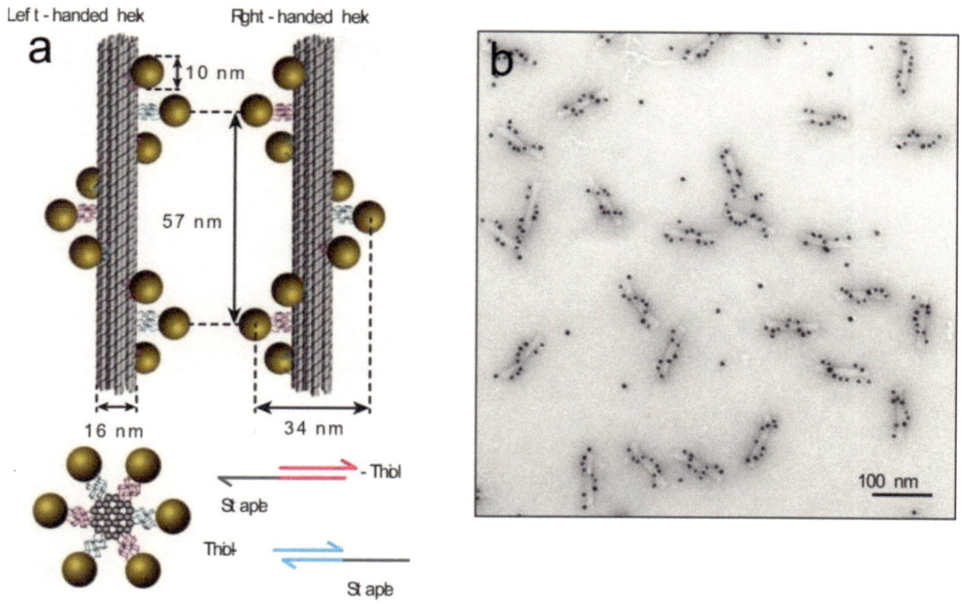

Figure 5.
Chiral assemblies of gold nanoparticles and DNA origami in left- and right-handed nanohelices: (a) schematics, (b) TEM image. Reproduced with permission from Kuzyk et al. [53], Copyright 2012 Springer Nature.

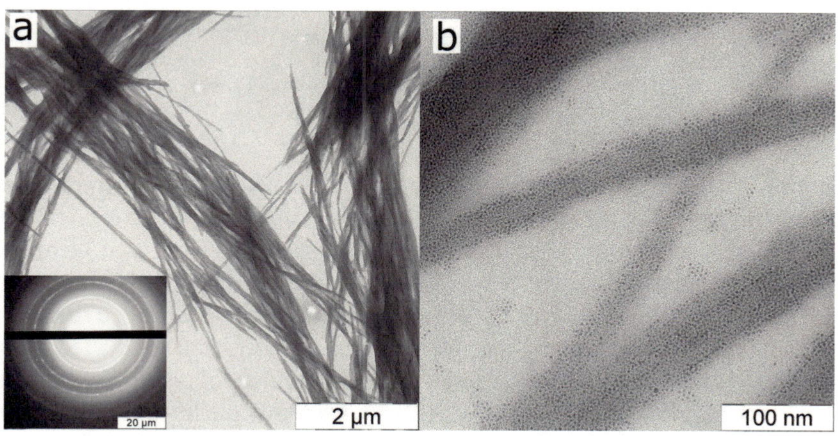

Figure 6.
TEM images of silver nanoparticle aggregates stabilized by thiocholesterol (after 24 h storage in mesophase): (a) general view, (b) internal structure, and (c) the formation scheme of silver-thiocholesterol chiral 3D-aggregates. Reproduced with permission from Shabatina et al. [62], Copyright 2013 Springer.

of them was thermodynamically more favorable. Glutathione molecules supported by cetylammonium bromide micelles directed the self-assembly of gold nanorods and the formation of chiral nanochains with end-to-end contacts [67].

A strong chiroptical signal can be generated by plasmon nanorods assembled into a three-dimensional spiral structure by supramolecular fibers, such as anthraquinone based oxalamide [68]. The DNA origami approach allowed you to adjust the chiro-optical response by carefully controlling of the plasmonic superstructure. Dynamic switching between different configurations of the nanorod spirals led to a change in the chiro-optical response [69].

Other biomolecular matrices also can be used to order plasmonic nanoparticles in three-dimensional chiral assemblies, such as proteins or their aggregates. So, it

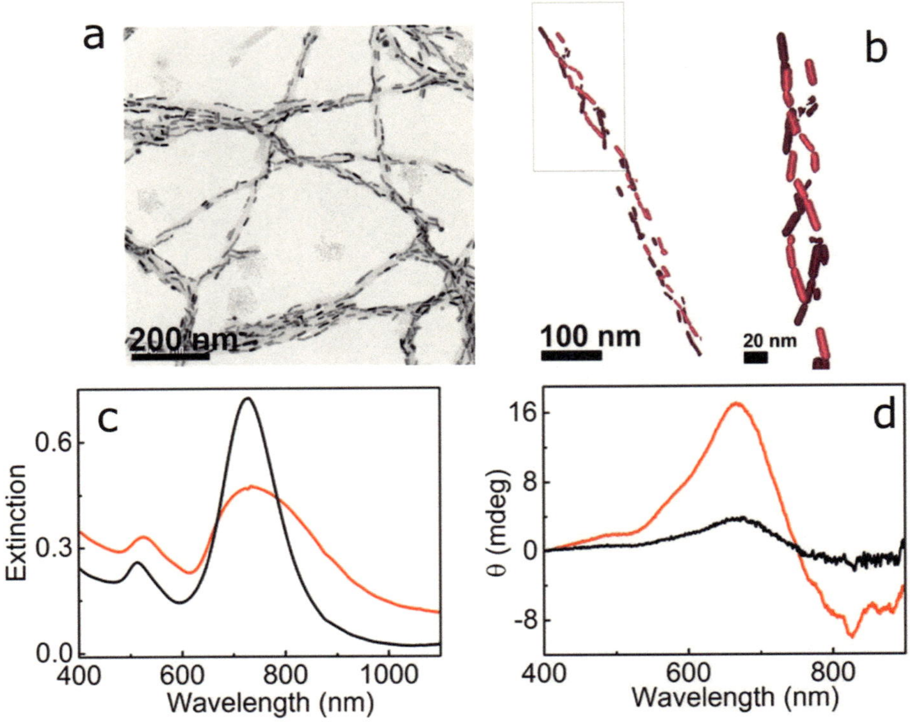

Figure 7.
Composite α-synuclein fiber with 3D chiral arrangement of Au nanorods: (a) TEM image, (b) cryo-TEM tomography reconstruction, (c) extinction, and (d) CD spectra of Au nanorods monitored 30 min after the addition of 30 μL of purified brain homogenates from healthy (black) and Parkinson-disease-affected (red) patients. Adapted with permission from Kumar et al. [60], Copyright 2018 National Academy of Sciences.

was reported that amyloid fibrils of α-synuclein CD were detected using helically arranged gold nanorods (see **Figure 7**) [60]. At the same time, chiro-optical activity was not detected when only α-synuclein monomers were present. This technique can be further expanded to detect infectious recombinant prions.

There are cases of aggregation of plasmonic nanoparticles, leading to the sharp optical changes upon initiation by only one enantiomer. D-glutamic acid led to the aggregation of gold nanoparticles coated with CTAB, which followed by a significant change in color, while for the L-enantiomer there were practically no changes [70]. Similarly, the enantioselective detection of D-cysteine by silver nanoparticles was recently demonstrated in the solution and in the special bacterial cellulose matrix [71].

5. Biosensing with 3D chiral nanostructures

To amplify the CD signal of a chiral molecule, the differential absorption of the chiral molecule must be increased. For an analytical estimate of the speed of light absorption, one can use the relation [72]:

$$A \propto \frac{\omega}{2}\left(\alpha''\left|\tilde{E}^*\right|^2 + \chi''\left|\tilde{B}\right|^2\right) - G''\omega Im\left(\tilde{E}^* \cdot \tilde{B}\right) \quad (1)$$

where ω is the angular frequency of electromagnetic wave, α'', χ'' and G'' are the imaginary parts of electric polarizability, magnetic susceptibility and mixed electric-magnetic polarizability of the molecule, and $\tilde{E}$ and $\tilde{B}$ are the complex electric and magnetic field, respectively. CD is determined by the second term, which is defined by both the intrinsic chirality of a molecule ($G'' \neq 0$) and by local electromagnetic field ($\mathrm{Im}\left(\tilde{E}^{*}\cdot\tilde{B}\right)$), which needs to have nonorthogonal components of electric and magnetic field vectors.

Thereby, to enhance the CD response of a chiral molecule one can modify the electromagnetic field to maximize the local optical chirality C, the quantity originally introduced in 1964 [73] and recently revised [72].

$$C = -\frac{\varepsilon_0 \omega}{2}\mathrm{Im}\left(\tilde{E}^{*}\cdot\tilde{B}\right) \tag{2}$$

The present subsections consider the research directed at constructing plasmonic systems with enhanced near-field chirality and examples of biosensing based on these nanostructures.

The efforts of many groups of scientists have been expended in order to increase the optical chirality C (Eq. (2)) using chiral plasmonic nanostructures. One of the chiral structures that scientists have studied both theoretically and experimentally is nanohelix. The numerical modeling of such structures determined a strong CD effect [74]. However, for gold nanohelices 400 nm high, 400 nm in diameter, and 80 nm thick nanowires, the chirality enhancement factor was determined as 10, so the use of such structures in biosensing is limited [75]. This problem was solved by fabricating superstructures with several nanohelices [76] using direct laser recording methods [77, 78] or ion- and electron beam lithography [79–81]. The use of such structures made it possible to increase the optical chiral amplification factor up to two orders of magnitude and to adjust the CD signal from IR to the visible region of the spectrum. A schematic representation of the structure and the field distribution are shown in **Figure 8**. In addition, this approach made it possible to create core-shell nanohelices with adjustable dissymmetry coefficients and CD [82].

Recently, another method using three-dimensional chiral plasmonic nanohelices for probing was demonstrated [83]. Instead of using enhanced near-field chirality, the CD signal of Pd nanohelices was used to detect a (0.1–1%) hydrogen content. Large-area arrays of Pd and Pd-Au hybrid nanostructures were obtained using the angular nanoglancing deposition method [84], which combines block copolymer micellar nanolithography [85] with glancing-angle deposition (GLAD) [86].

The last considered approach for synthesis of chiral lattice is adsorption of chiral organic molecules on growing nanocrystals. The following review demonstrated

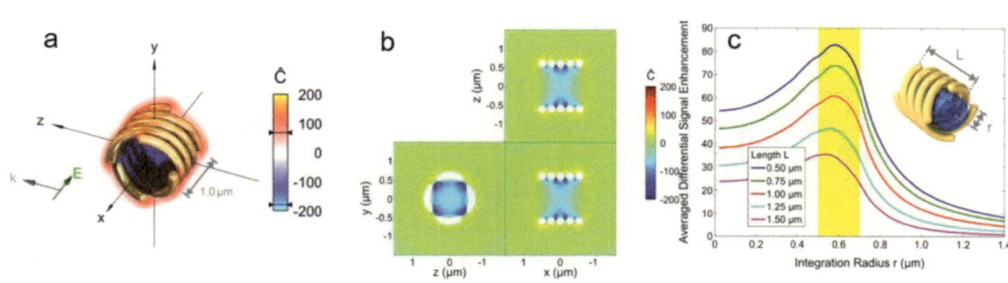

Figure 8.
3D chiral plasmonic nanostructures: (a, b) simulated four-helix gold nanostructure showing strongly confined optical chirality within the helices and (c) the graph presenting the enhancement of differential absorption signal depending on the radius of integration, for different length of a structure. Adapted with permission from Schäferling et al. [76], Copyright 2014 American Chemical Society.

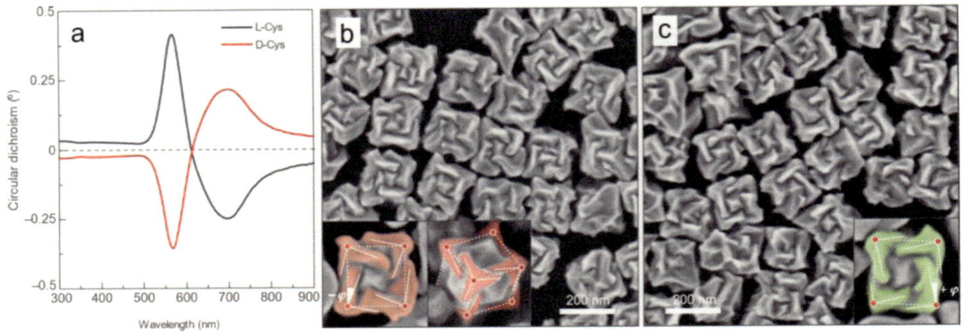

Figure 9.
Three-dimensional plasmonic helicoids controlled by cysteine chirality transfer: (a) CD spectra and (b–c) SEM images of chiral nanoparticles synthesized using L-Cys and D-Cys. The insets highlight the tilted edges (solid lines), cubic outline (dashed lines), and tilt angles (−φ and +φ). Reproduced with permission from Lee et al. [77], Copyright 2018 Springer Nature.

the formation of chiral nanocrystals of Te and Se, which can be used as a matrix for growing nanostructures of gold and silver tellurides [87]. Also, there was presented the growth of chiral gold nanoparticles in the opposite direction induced by amino acids and peptides [77]. The effect of different growth rates on the chiral morphology of gold nanocrystals in the presence of L-Cys or D-Cys was observed. The structure and CD spectra of these nanocrystals are shown in **Figure 9**. It has been revealed that in the presence of L-glutathione (GSH) growing nanocrystals have different morphologies. The proposed mechanism involves specific adsorption of Cys or GSH on the high-index planes of growing particles.

6. Planar chiral nanostructures and metasurfaces

Next perspective structures are planar plasmonic nanostructures with thickness less than the wavelength of the incident light. They are very attractive from a technological point of view because of a possibility of mass production using conventional lithographic methods. As well as for truly chiral forms with enantiomers that cannot be superimposed by any rotation in 3D space, the left and right enantiomers of a flat chiral shape are not superimposed on each other by the rotation in a plane. Handling in the latter case has been determined by the viewing side, which leads to inverted CD spectra when the planar structure is illuminated from the opposite normal direction. In contrast to three-dimensional spiral structures, the differential light absorption of LCP and RCP by planar plasmon nanostructures is due to the difference in LSPR-induced near-field distributions [88]. Similar effects were observed in the study of 2D chiral plasmon nanostructures (see **Figure 10**), such as G- [89], S- [90], L-shaped [91] nanostructures, gammadions [92, 93], nanohelix [75] asymmetric nanoparticles [94], checkerboard nanorods [95] and also for flat metamaterials, including thin metal films with two-dimensional chiral holes, such as dimers with nanogap [96, 97].

Chiroptical response of planar chiral plasmonic nanostructures can be highly sensitive to the presence and to the specifics of tertiary structure of biomolecules, as in the case of arrays of gold gammadion nanostructures [98]. These effects can be viewed in plasmonic peak shifts for RCP and LCP light upon adsorption of proteins and in the difference of the values $\Delta\Delta\lambda = \Delta\lambda RCP - \Delta\lambda LCP$, and can been used as an analytic signal. The parameter $\Delta\Delta\lambda$ becomes zero: $\Delta\Delta\lambda = 0$ for achiral adsorbed molecules. In this way, the picogram levels of proteins were detected.

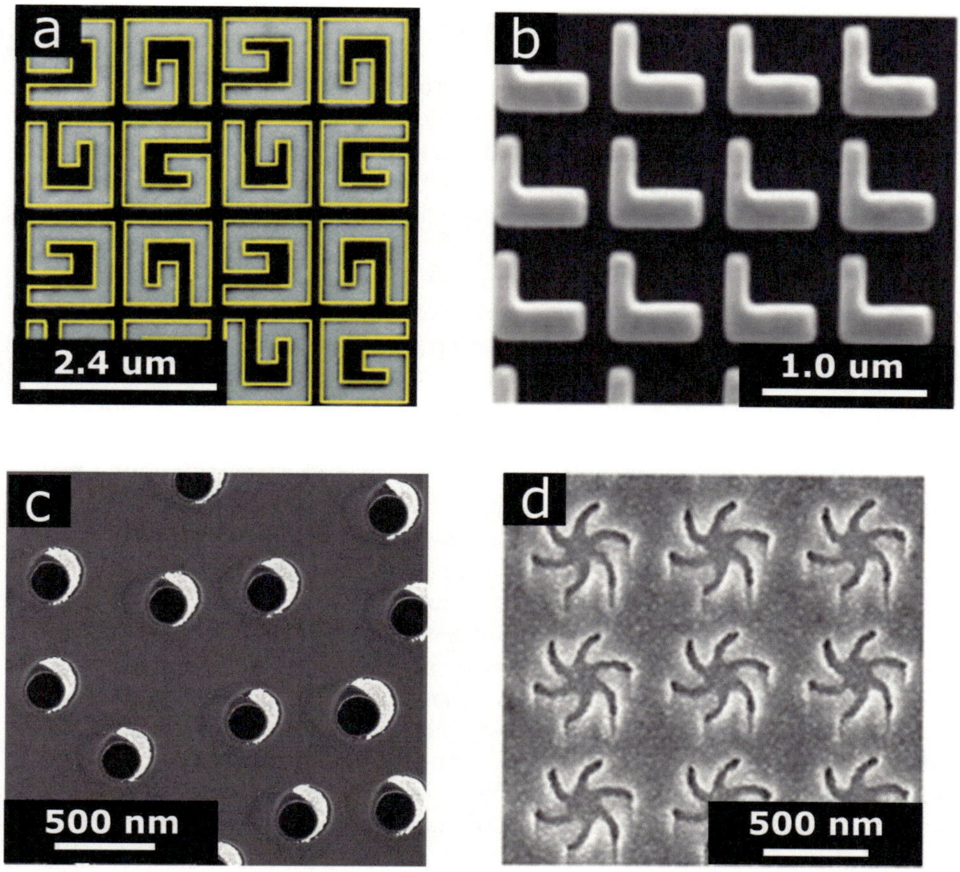

Figure 10.
Examples of planar chiral nanostructures: (a) arrays of G-shaped gold nanoparticles. Adapted with permission from Valev et al. [89], Copyright 2009 American Chemical Society. (b) Arrays of L-shaped gold nanoparticles. Adapted with permission from Ye et al. [91], Copyright 2017 American Physical Society. (c) Short-ordered arrays of comma-shaped gold nanoparticles [94]. (d) Gold metafilm with arrays of "shuriken" nanostructures. Adapted with permission from Kelly et al. [99], Copyright 2018 American Chemical Society.

Furthermore, the proteins with different content of β-sheets give different CD response, so, providing information about the structure of the adsorbed biomolecules. This feature has been studied in detail using chiral "shuriken"-like gold-covered indentations created on polycarbonate templates using injection molding method and gold deposition [100]. The chiroptic properties of the far-field were characterized by the collection of ORD spectra in the reflection mode for linearly polarized incident light. The observed peak shifts demonstrated picogram limit of detection of protein. Conformational changes associated with protein binding to ligands can be accompanied by the changes in the asymmetry factor $\Delta\Delta\lambda$, demonstrating sensitivity to the tertiary and domain (quaternary) structure of proteins [101].

More recent studies have shown that these chiral plasmon substrates can distinguish between proteins that have similar structures but have primary sequences that differ in one amino acid [102]. So, it can provide information about the structural order in complex biointerfaces [99].

Thus, the "superchiral" fields generated by 2D chiral nanostructures can provide a unique opportunity to probe the chirality of adsorbed biomolecules at such levels, as conformation, orientation, molecular structure, and supramolecular ordering.

7. Conclusions

Unique properties of plasmonic nanostructures open exciting new possibilities for increasing the sensitivity of molecular chirality. There are several common approaches among different experimental and theoretical studies benefiting in biosensing. Despite moderate CD enhancement factors, the formation of complexes with small achiral nanoparticles can be useful for detecting chiral organic molecules and peptides.

An alternative approach based on the formation/destruction of chiral plasmon dimers, constructed using target-specific aptamers, makes it possible to detect a wide range of biomolecules, including enzymes and proteins, in complex biological environments and even inside living cells. For detection of supramolecular biological structures, such as amyloid fibrils, the formation of self-organizing three-dimensional chiral assemblies of achiral plasmon nanoparticles can be used. This fact is very useful for early diagnosis of various concomitant diseases. Sophisticated fabrication and a moderate increase in optical chirality make it difficult to use three-dimensional chiral plasmon nanostructures, while flat chiral metasurfaces that can be conveniently obtained using lithographic methods demonstrate a high potential for ex-vivo detection of chiral biomolecules, further allowing the study of their tertiary and domain structure.

Currently, the field of chiral plasmonic biosensing is of great interest for various biosensor applications [103]. One can expect new developments in this area and unique results in the future. Due to the theoretical basis developed by scientists in understanding the nature of the interaction of chiral molecules with plasmon-induced near-fields, the creation of more advanced nanostructures requires even greater care. Variation of the geometry and assembly of nanostructures, as well as the use of new, primarily dielectric [104] and hybrid materials and alloys with specified dielectric functions [105]—these are the ways of increasing sensitivity of chiral biosensors. Moreover, the use of combinations of various types of chiral nanostructures allows the simultaneous detection of chirality at various scales or the detection of various chiral molecules in the complex biological media.

Acknowledgements

The authors thank Russian Foundation of Basic Research, grant no. 18-03-00730a.

Author details

Vladimir E. Bochenkov* and Tatyana I. Shabatina*
Department of Chemistry, M.V. Lomonosov Moscow State University, Moscow, Russian Federation

*Address all correspondence to: boch@kinet.chem.msu.ru and tatyanashabatina@yandex.ru

References

[1] Mayer KM, Hafner JH. Localized surface plasmon resonance sensors. Chemical Reviews. 2011;**111**:3828-3857

[2] Hammond JL, Bhalla N, Rafiee SD, Estrela P. Localized surface plasmon resonance as a biosensing platform for developing countries. Biosensors. 2014;**4**:172-188

[3] Li M, Cushing SK, Wu N. Plasmon-enhanced optical sensors: A review. Analyst. 2015;**140**:386-406

[4] Pesek JJ, Matyska MT, Pik Fong F. Evaluation of the silanization/hydrosilation process for the synthesis of chiral stationary phases. Chromatographia. 2001;**53**:635

[5] Grzywinski D, Szumski M, Buszewski B. Polymer monoliths with silver nanoparticles-cholesterol conjugate as stationary phases for capillary liquid chromatography. Journal of Chromatography. A. 2017;**1526**:2-11

[6] Lammerhofer M. Chiral recognition by enantioselective liquid chromatography: Mechanisms and modern chiral stationary phases. Journal of Chromatography. A. 2010;**1217**:814-856

[7] Schäferling M. Chiral Nanophotonics; Springer Series in Optical Sciences. Vol. 205. Cham, Switzerland: Springer International Publishing; 2017. p. 159

[8] Hentschel M, Schäferling M, Duan X, Giessen H, Liu N. Chiral plasmonics. Science Advances. 2017;**3**:e1602735

[9] Wang X, Tang Z. Circular dichroism studies on plasmonic nanostructures. Small. 2017;**13**:1601115

[10] Ben-Moshe A, Maoz BM, Govorov AO, Markovich G. Chirality and chiroptical effects in inorganic nanocrystal systems with plasmon and exciton resonances. Chemical Society Reviews. 2013;**42**:7028-7041

[11] Valev VK, Baumberg JJ, Sibilia C, Verbiest T. Chirality and chiroptical effects in plasmonic nanostructures: Fundamentals, recent progress, and outlook. Advanced Materials. 2013;**25**:2517-2534

[12] Collins JT, Kuppe C, Hooper DC, Sibilia C, Centini M, Valev VK. Chirality and chiroptical effects in metal nanostructures: Fundamentals and current trends. Advanced Optical Materials. 2017;**5**:1700182

[13] Aizpurua J, Bryant GW, Richter LJ, García De Abajo FJ, Kelley BK, Mallouk T. Optical properties of coupled metallic nanorods for field-enhanced spectroscopy. Physical Review B: Condensed Matter and Materials Physics. 2005;**71**:235420

[14] Lieberman I, Shemer G, Fried T, Kosower E, Markovich G. Plasmon-resonance-enhanced absorption and circular Dichroism. Angewandte Chemie, International Edition. 2008;**47**:4855-4857

[15] Govorov AO, Fan Z, Hernandez P, Slocik JM, Naik RR. Theory of circular dichroism of nanomaterials comprising chiral molecules and nanocrystals: Plasmon enhancement, dipole interactions, and dielectric effects. Nano Letters. 2010;**10**:1374-1382

[16] García-Etxarri A, Dionne JA. Surface-enhanced circular dichroism spectroscopy mediated by nonchiral nanoantennas. Physical Review B. 2013;**87**:235409

[17] Yoo S, Park QH. Enhancement of chiroptical signals by circular differential Mie scattering of

nanoparticles. Scientific Reports. 2015;**5**:14463

[18] Chulhai DV, Jensen L. Plasmonic circular dichroism of 310- and α-helix using a discrete interaction model/quantum mechanics method. The Journal of Physical Chemistry. A. 2015;**119**:5218-5223

[19] Slocik JM, Govorov AO, Naik RR. Plasmonic circular dichroism of peptide-functionalized gold nanoparticles. Nano Letters. 2011;**11**:701-705

[20] Maoz BM, Chaikin Y, Tesler AB, Bar Elli O, Fan Z, Govorov AO, et al. Amplification of chiroptical activity of chiral biomolecules by surface plasmons. Nano Letters. 2013;**13**:1203-1209

[21] Rezanka P, Záruba K, Král V. Supramolecular chirality of cysteine modified silver nanoparticles. Colloids and Surfaces A: Physicochemical and Engineering Aspects. 2011;**374**:77-83

[22] di Gregorio MC, Ben Moshe A, Tirosh E, Galantini L, Markovich G. Chiroptical study of plasmon–molecule interaction: The case of interaction of glutathione with silver nanocubes. Journal of Physical Chemistry C. 2015;**119**:17111-17116

[23] Kobayashi M, Tomita S, Sawada K, Shiba K, Yanagi H, Yamashita I, et al. Chiral meta-molecules consisting of gold nanoparticles and genetically engineered tobacco mosaic virus. Optics Express. 2012;**20**:24856

[24] Lu F, Tian Y, Liu M, Su D, Zhang H, Govorov AO, et al. Discrete nanocubes as plasmonic reporters of molecular chirality. Nano Letters. 2013;**13**:3145-3151

[25] McPeak KM, van Engers CD, Bianchi S, Rossinelli A, Poulikakos LV, Bernard L, et al. Ultraviolet plasmonic chirality from colloidal aluminum nanoparticles exhibiting charge-selective protein detection. Advanced Materials. 2015;**27**:6244-6250

[26] Levi-Belenkova T, Govorov AO, Markovich G. Orientation-sensitive peptide-induced plasmonic circular dichroism in silver nanocubes. Journal of Physical Chemistry C. 2016;**120**:12751-12756

[27] Govorov AO. Plasmon-induced circular dichroism of a chiral molecule in the vicinity of metal nanocrystals. Application to various geometries. Journal of Physical Chemistry C. 2011;**115**:7914-7923

[28] Zhang H, Govorov AO. Giant circular dichroism of a molecule in a region of strong plasmon resonances between two neighboring gold nanocrystals. Physical Review B. 2013;**87**:075410

[29] Tian X, Fang Y, Sun M. Formation of enhanced uniform chiral fields in symmetric dimer nanostructures. Scientific Reports. 2015;**5**:17534

[30] Besteiro LV, Zhang H, Plain J, Markovich G, Wang Z, Govorov AO. Aluminum nanoparticles with hot spots for plasmon-induced circular dichroism of chiral molecules in the UV spectral interval. Advanced Optical Materials. 2017;**5**:1700069

[31] Nesterov ML, Yin X, Schäferling M, Giessen H, Weiss T. The role of plasmon-generated near fields for enhanced circular dichroism spectroscopy. ACS Photonics. 2016;**3**:578-583

[32] Wu X, Xu L, Liu L, Ma W, Yin H, Kuang H, et al. Unexpected chirality of nanoparticle dimers and ultrasensitive chiroplasmonic bioanalysis. Journal of the American Chemical Society. 2013;**135**:18629-18636

[33] Gérard VA, Gun'ko YK, Defrancq E, Govorov AO. Plasmon-induced CD response of oligonucleotide-conjugated metal nanoparticles. Chemical Communications. 2011;**47**:7383-7385

[34] Hao C, Xu L, Ma W, Wang L, Kuang H, Xu C. Assembled plasmonic asymmetric heterodimers with tailorable chiroptical response. Small. 2014;**10**:1805-1812

[35] Zhu Z, Liu W, Li Z, Han B, Zhou Y, Gao Y, et al. Manipulation of collective optical activity in one-dimensional plasmonic assembly. ACS Nano. 2012;**6**:2326-2332

[36] Zhu F, Li X, Li Y, Yan M, Liu S. Enantioselective circular dichroism sensing of cysteine and glutathione with gold nanorods. Analytical Chemistry. 2015;**87**:357-361

[37] Xu L, Xu Z, Ma W, Liu L, Wang L, Kuang H, et al. Highly selective recognition and ultrasensitive quantification of enantiomers. Journal of Materials Chemistry B. 2013;**1**:4478-4483

[38] Wang RY, Wang P, Liu Y, Zhao W, Zhai D, Hong X, et al. Experimental observation of giant chiroptical amplification of small chiral molecules by gold nanosphere clusters. Journal of Physical Chemistry C. 2014;**118**:9690-9695

[39] Layani ME, Ben Moshe A, Varenik M, Regev O, Zhang H, Govorov AO, et al. Chiroptical activity in silver cholate nanostructures induced by the formation of nanoparticle assemblies. Journal of Physical Chemistry C. 2013;**117**:22240-22244

[40] Xu Z, Xu L, Zhu Y, Ma W, Kuang H, Wang L, et al. Chirality based sensor for bisphenol A detection. Chemical Communications. 2012;**48**:5760-5762

[41] Xu Z, Xu L, Liz-Marzán LM, Ma W, Kotov NA, Wang L, et al. Sensitive detection of silver ions based on chiroplasmonic assemblies of nanoparticles. Advanced Optical Materials. 2013;**1**:626-630

[42] Liu Y, Wei M, Zhang L, Wei W, Zhang Y, Liu S. Evaluation of DNA methyltransferase activity and inhibition via chiroplasmonic assemblies of gold nanoparticles. Chemical Communications. 2015;**51**:14350-14353

[43] Zhao X, Wu X, Xu L, Ma W, Kuang H, Wang L, et al. Building heterogeneous core–satellite chiral assemblies for ultrasensitive toxin detection. Biosensors & Bioelectronics. 2015;**66**:554-558

[44] Zhao Y, Xu L, Ma W, Wang L, Kuang H, Xu C, et al. Shell-engineered chiroplasmonic assemblies of nanoparticles for zeptomolar DNA detection. Nano Letters. 2014;**14**:3908-3913

[45] Zhao H, Bian S, Yang Y, Wu X. Chiroplasmonic assemblies of gold nanoparticles as a novel method for sensitive detection of alpha-fetoprotein. Microchimica Acta. 2017;**184**:1855-1862

[46] Liu Y, Wei M, Zhang L, Zhang Y, Wei W, Yin L, et al. Chiroplasmonic assemblies of gold nanoparticles for ultrasensitive detection of 8-hydroxy-2′-deoxyguanosine in human serum sample. Analytical Chemistry. 2016;**88**:6509-6514

[47] Fu P, Sun M, Xu L, Wu X, Liu L, Kuang H, et al. A self-assembled chiral-aptasensor for ATP activity detection. Nanoscale. 2016;**8**:15008-15015

[48] Sun M, Xu L, Fu P, Wu X, Kuang H, Liu L, et al. Scissor-like chiral metamolecules for probing intracellular telomerase activity. Advanced Functional Materials. 2016;**26**:7352-7358

[49] Kneer LM, Roller EM, Besteiro LV, Schreiber R, Govorov AO, Liedl T. Circular dichroism of chiral molecules

in DNA-assembled plasmonic hotspots. ACS Nano. 2018;**12**:9110-9115

[50] Lan X, Wang Q. Self-assembly of chiral plasmonic nanostructures. Advanced Materials. 2016;**28**:10499-10507

[51] Fan Z, Govorov AO. Plasmonic circular dichroism of chiral metal nanoparticle assemblies. Nano Letters. 2010;**10**:2580-2587

[52] Fan Z, Govorov AO. Helical metal nanoparticle assemblies with defects: Plasmonic chirality and circular dichroism. Journal of Physical Chemistry C. 2011;**115**:13254-13261

[53] Kuzyk A, Schreiber R, Fan Z, Pardatscher G, Roller EM, Högele A, et al. DNA-based self-assembly of chiral plasmonic nanostructures with tailored optical response. Nature. 2012;**483**:311-314

[54] Zhao Y, Xu L, Kuang H, Wang L, Xu C. Asymmetric and symmetric PCR of gold nanoparticles: A pathway to scaled-up self-assembly with tunable chirality. Journal of Materials Chemistry. 2012;**22**:5574-5580

[55] Macfarlane RJ, Lee B, Jones MR, Harris N, Schatz GC, Mirkin CA. Nanoparticle superlattice engineering with DNA. Science. 2011;**334**:204-208

[56] Cecconello A, Besteiro LV, Govorov AO, Willner I. Chiroplasmonic DNA-based nanostructures. Nature Reviews Materials. 2017;**2**:17039

[57] Merg AD, Boatz JC, Mandal A, Zhao G, Mokashi-Punekar S, Liu C, et al. Peptide-directed assembly of single-helical gold nanoparticle superstructures exhibiting intense chiroptical activity. Journal of the American Chemical Society. 2016;**138**:13655-13663

[58] Mokashi-Punekar S, Merg AD, Rosi NL. Systematic adjustment of pitch and particle dimensions within a family of chiral plasmonic gold nanoparticle single helices. Journal of the American Chemical Society. 2017;**139**:15043-15048

[59] Song C, Blaber MG, Zhao G, Zhang P, Fry HC, Schatz GC, et al. Tailorable plasmonic circular dichroism properties of helical nanoparticle superstructures. Nano Letters. 2013;**13**:3256-3261

[60] Kumar J, Eraña H, López-Martínez E, Claes N, Martín VF, Solís DM, et al. Detection of amyloid fibrils in Parkinson's disease using plasmonic chirality. Proceedings of the National Academy of Sciences of the United States of America. 2018;**115**:3225-3230

[61] Shabatina TI, Belyaev AA, Sergeev GB. Silver/thiocholesterol and silver/cholesterol nanosized aggregates formation in liquid crystalline mesophase. Molecular Crystals and Liquid Crystals. 2011;**540**:169-174

[62] Shabatina TI, Belyaev AA, Sergeev GB. Self-assembled nanostructures in silver–cholesterol and silver–thiocholesterol systems. BioNanoScience. 2013;**3**:289-294

[63] Shabatina TI, Gromova YA, Anistratova ES, Belyaev AA. New chiral metal-mesogenic nanosystems "silver–thiocholesterol" and their adsorption properties. Molecular Crystals and Liquid Crystals. 2016;**632**:64-69

[64] Auguié B, Alonso-Gómez JL, Guerrero-Martínez A, Liz-Marzán LM. Fingers crossed: Optical activity of a chiral dimer of plasmonic nanorods. Journal of Physical Chemistry Letters. 2011;**2**:846-851

[65] Ma W, Kuang H, Wang L, Xu L, Chang WS, Zhang H, et al. Chiral

plasmonics of self-assembled nanorod dimers. Scientific Reports. 2013;**3**:1934

[66] Ma W, Kuang H, Xu L, Ding L, Xu C, Wang L, et al. Attomolar DNA detection with chiral nanorod assemblies. Nature Communications. 2013;**4**:2689

[67] Lu J, Chang YX, Zhang NN, Wei Y, Li AJ, Tai J, et al. Chiral plasmonic nanochains via the self-assembly of gold nanorods and helical glutathione oligomers facilitated by cetyltrimethylammonium bromide micelles. ACS Nano. 2017;**11**:3463-3475

[68] Guerrero-Martínez A, Auguié B, Alonso-Gómez JL, Džolić Z, Gómez-Graña S, Žinić M, et al. Intense optical activity from three-dimensional chiral ordering of plasmonic nanoantennas. Angewandte Chemie, International Edition. 2011;**50**:5499-5503

[69] Lan X, Liu T, Wang Z, Govorov AO, Yan H, Liu Y. DNA-guided plasmonic helix with switchable chirality. Journal of the American Chemical Society. 2018;**140**:11763-11770

[70] Wang Y, Zhou X, Xu C, Jin Y, Li B. Gold nanorods as visual sensing platform for chiral recognition with naked eyes. Scientific Reports. 2018;**8**:5296

[71] Zor E. Silver nanoparticles-embedded nanopaper as a colorimetric chiral sensing platform. Talanta. 2018;**184**:149-155

[72] Tang Y, Cohen AE. Optical chirality and its interaction with matter. Physical Review Letters. 2010;**104**:163901

[73] Lipkin DM. Existence of a new conservation law in electromagnetic theory. Journal of Mathematical Physics. 1964;**5**:696-700

[74] Gansel JK, Wegener M, Burger S, Linden S. Gold helix photonic metamaterials: A numerical parameter study. Optics Express. 2010;**18**:1059-1069

[75] Schäferling M, Dregely D, Hentschel M, Giessen H. Tailoring enhanced optical chirality: Design principles for chiral plasmonic nanostructures. Physical Review X. 2012;**2**:031010

[76] Schäferling M, Yin X, Engheta N, Giessen H. Helical plasmonic nanostructures as prototypical chiral near-field sources. ACS Photonics. 2014;**1**:530-537

[77] Lee HE, Ahn HY, Mun J, Lee YY, Kim M, Cho NH, et al. Amino-acidand peptide-directed synthesis of chiral plasmonic gold nanoparticles. Nature. 2018;**556**:360-365

[78] Gansel JK, Thiel M, Rill MS, Decker M, Bade K, Saile V, et al. Gold helix photonic metamaterial as broadband circular polarizer. Science. 2009;**325**:1513-1515

[79] Esposito M, Tasco V, Todisco F, Benedetti A, Sanvitto D, Passaseo A. Three dimensional chiral metamaterial nanospirals in the visible range by vertically compensated focused ion beam induced-deposition. Advanced Optical Materials. 2014;**2**:154-161

[80] Esposito M, Tasco V, Cuscunà M, Todisco F, Benedetti A, Tarantini I, et al. Nanoscale 3D chiral plasmonic helices with circular dichroism at visible frequencies. ACS Photonics. 2015;**2**:105-114

[81] Esposito M, Tasco V, Todisco F, Cuscunà M, Benedetti A, Scuderi M, et al. Programmable extreme chirality in the visible by helix-shaped metamaterial platform. Nano Letters. 2016;**16**:5823-5828

[82] Kosters D, de Hoogh A, Zeijlemaker H, Acar H, Rotenberg N, Kuipers L. Core–shell plasmonic nanohelices. ACS Photonics. 2017;**4**:1858-1863

[83] Matuschek M, Singh DP, Jeong HHH, Nesterov M, Weiss T, Fischer P, et al. Chiral plasmonic hydrogen sensors. Small. 2018;**14**:1702990

[84] Mark AG, Gibbs JG, Lee TC, Fischer P. Hybrid nanocolloids with programmed three-dimensional shape and material composition. Nature Materials. 2013;**12**:802-807

[85] Glass R, Möller M, Spatz JP. Block copolymer micelle nanolithography. Nanotechnology. 2003;**14**:1153-1160

[86] Abbas F, Faryad M. A highly sensitive multiplasmonic sensor using hyperbolic chiral sculptured thin films. Journal of Applied Physics. 2017;**122**:173104

[87] Ben-Moshe A, Wolf SG, Sadan MB, Houben L, Fan Z, Govorov AO, et al. Enantioselective control of lattice and shape chirality in inorganic nanostructures using chiral biomolecules. Nature Communications. 2014;**5**:4302

[88] Kong XT, Besteiro LV, Wang Z, Govorov AO. Plasmonic chirality and circular dichroism in bioassembled and nonbiological systems: Theoretical background and recent progress. Advanced Materials. 2018;(1801790). DOI: 10.1002/adma.201801790

[89] Valev VK, Smisdom N, Silhanek AV, De Clercq B, Gillijns W, Ameloot M, et al. Plasmonic ratchet wheels: Switching circular dichroism by arranging chiral nanostructures. Nano Letters. 2009;**9**:3945-3948

[90] Narushima T, Okamoto H. Circular dichroism nano-imaging of two-dimensional chiral metal nanostructures. Physical Chemistry Chemical Physics. 2013;**15**:13805-13809

[91] Ye W, Yuan X, Guo C, Zhang J, Yang B, Zhang S. Large chiroptical effects in planar chiral metamaterials. Physical Review Applied. 2017;**7**:054003

[92] Papakostas A, Potts A, Bagnall DM, Prosvirnin SL, Coles HJ, Zheludev NI. Optical manifestations of planar chirality. Physical Review Letters. 2003;**90**:107404

[93] Kuwata-Gonokami M, Saito N, Ino Y, Kauranen M, Jefimovs K, Vallius T, et al. Giant optical activity in quasi-two-dimensional planar nanostructures. Physical Review Letters. 2005;**95**:227401

[94] Bochenkov VE, Sutherland DS. Chiral plasmonic nanocrescents: Large-area fabrication and optical properties. Optics Express. 2018;**26**:27101-27108

[95] Meinzer N, Hendry E, Barnes WL. Probing the chiral nature of electromagnetic fields surrounding plasmonic nanostructures. Physical Review B. 2013;**88**:041407

[96] Hendry E, Mikhaylovskiy RV, Barron LD, Kadodwala M, Davis TJ. Chiral electromagnetic fields generated by arrays of nanoslits. Nano Letters. 2012;**12**:3640-3644

[97] Zhukovsky SV, Chigrin DN, Kremers C, Lavrinenko AV. Dichroism, chirality, and polarization eigenstates in Babinet nanoslot-dimer membrane metamaterials. Photonics and Nanostructures - Fundamentals and Applications. 2013;**11**:353-361

[98] Hendry E, Carpy T, Johnston J, Popland M, Mikhaylovskiy RV, Lapthorn AJ, et al. Ultrasensitive detection and characterization of biomolecules using superchiral fields. Nature Nanotechnology. 2010;**5**:783-787

[99] Kelly C, Tullius R, Lapthorn AJ, Gadegaard N, Cooke G, Barron LD, et al. Chiral plasmonic fields probe structural order of biointerfaces. Journal of the American Chemical Society. 2018;**140**:8509-8517

[100] Karimullah AS, Jack C, Tullius R, Rotello VM, Cooke G, Gadegaard N, et al. Disposable plasmonics: Plastic templated plasmonic metamaterials with tunable chirality. Advanced Materials. 2015;**27**:5610-5616

[101] Tullius R, Karimullah AS, Rodier M, Fitzpatrick B, Gadegaard N, Barron LD, et al. "Superchiral" spectroscopy: Detection of protein higher order hierarchical structure with chiral plasmonic nanostructures. Journal of the American Chemical Society. 2015;**137**:8380-8383

[102] Tullius R, Platt GW, Khosravi Khorashad L, Gadegaard N, Lapthorn AJ, Rotello VM, et al. Superchiral plasmonic phase sensitivity for fingerprinting of protein interface structure. ACS Nano. 2017;**11**:12049-12056

[103] Bochenkov VE, Shabatina TI. Chiral plasmonic biosensors. Biosensors. 2018;**8**(4):120

[104] Mohammadi E, Tsakmakidis KL, Askarpour AN, Dehkhoda P, Tavakoli A, Altug H. Nanophotonic platforms for enhanced chiral sensing. ACS Photonics. 2018;**5**:2669-2675

[105] Jeong HH, Mark AG, Alarcón-Correa M, Kim I, Oswald P, Lee TC, et al. Dispersion and shape engineered plasmonic nanosensors. Nature Communications. 2016;**7**:11331

Chapter 4

Fullerene Based Sensor and Biosensor Technologies

Hilmiye Deniz Ertuğrul Uygun and Zihni Onur Uygun

Abstract

Sensor and biosensor technologies have shown rapid progress in recent years. These technologies use nanomaterials that have an important place in immobilization materials for recognition analyte molecules. Although fullerenes among these materials have attracted much attention in recent years, their number of studies is less than other carbon-based nanomaterials. Thanks to its completely closed structure and at least 30 double bonds, it can be modified from 30 points, which provides a great advantage. At these points, thanks to the ability to modify amine, thiol, carboxyl or metallic groups, modification residues can be created for all kinds of immobilization. According to the zero-dimensional nanomaterial class, fullerenes provide an extremely large surface area. Therefore, it provides more biological or non-biological recognition receptors immobilized on this surface area. Moreover, increasing the surface area with more recognition agent also increases the sensitivity. This is the most important parameter of sensor technologies, which is provided by fullerenes. In this book chapter, the development of fullerene-modified sensor and biosensor technologies are explained with examples, and fullerene modifications are given in figures as fullerene derivatives. Contribution was made in the method development stage by giving comparison of fullerene type sensor and biosensor systems.

Keywords: fullerene, sensor, biosensor, molecular imprinting, nanomaterials

1. Introduction

Carbon, which is the source of our lives, apart from our metabolic activities, attracts attention with its extraordinary structures created in nature. These structures are materials that are formed and discovered over time in the environment of high temperature and pressure [1]. The chemical properties of these materials are very different from those of inert carbon. Conductivity, strength, and catalytic properties are only a few of the carbon nanomaterials. By taking advantage of these features, the ability of today's technology to further develop products or make R&D increases. This, together with costs, can facilitate the development of technology.

With the development of biotechnology and the interdisciplinary sciences over time, the use of nanoscale materials is increasing in biotechnological process developments. Nanomaterials created new opportunities especially in biotechnology, with their easy modification advantage, especially in the field of diagnosis, and they offered significant advantages over traditional diagnostic methods in terms of sensitivity and selectivity. Diagnosis is the most important step in terms of developing health technologies. The correct diagnosis brings with it the rightful treatment, the right prognosis, the well-being of the patient, and the decrease in health expenditures. The important parameters in the exact diagnosis are sensitivity and accuracy. These two terms can describe the technological power of the diagnostic systems. In the development of sensitivity and accurate measurement, nanomaterials have an important place in today's technologies [2].

Among the various nanomaterials, carbon nanomaterials offer wide advantages due to their outstanding electrical, thermal, chemical, and mechanical properties [3]. Composite materials derived from carbon nanomaterials are used in energy storage and conversion, sensors, drug delivery, field emission, and nanoscale electronic components [4].

Depending on the purpose of use, carbon nanomaterials increase sensitivity by increasing surface area and conductivity especially in diagnostic systems. A promising sub-branch of diagnostic systems has made great progress in recent years, creating an important area in the development of point-of-care diagnostic tests. This area is especially developed on the fundamentals of sensor and biosensor technology. The technology consists of a recognition agent placed on a physicochemical transducer. In this simple system, electrodes, optic systems, or piezoelectric systems can be used as physicochemical transducers. Electrodes are physicochemical conductors that can detect electrochemical signals in a solution. On the other hand, optical sensors can detect light-matter interactions, and piezoelectric systems can perform specific and sensitive mass analysis. The recognition layer on these transducers plays a key role for biosensors and sensors. In biosensor systems, this recognition receptor is called biorecognition agent such as enzymes, antibodies, DNA, RNA, and other proteins that can be used as biorecognition elements [5]. As a result of the interaction of these biomolecules with the target molecule, catalytic or affinity-based biosensor systems can be developed. Otherwise, molecularly imprinted polymers, nanoparticles, and other polymers can be used as recognition agents in sensor systems instead of biological receptors [6]. Increasing the effectiveness of these recognition agents depends entirely on the properties of the immobilization/modification material used in the modification of the physicochemical transducer. Fortunately, nanomaterials can be used in sensor and biosensor systems in order to increase the power of the measurement system or to use it as recognition materials. These materials increase the surface area to obtain more sensitive signals and increase the possibility of interacting with more target molecules by binding more recognition agents to the surface. Technically, nanomaterial forms of inert metal/organic materials can be used as catalytic agents when they are in nano form.

In this book chapter, the production method, modification, and use of fullerene nanomaterials, which is a nanomaterial in the development of biosensor and sensor systems developed with biological or non-biological recognition agents, are described.

2. Fullerenes

2.1 Fullerene structure

Production and applications of carbon-based nanomaterials have gained speed in recent years. Carbon nanomaterials that can be found in different conformations with Sp2 hybridization can be extremely useful (**Figure 1**). These nanomaterials include nanotubes, graphene, carbon nanoparticles, carbon fibers, and fullerenes [7]. Among these materials, fullerenes are nanomaterials that have gained speed in recent years due to their structure and unique properties.

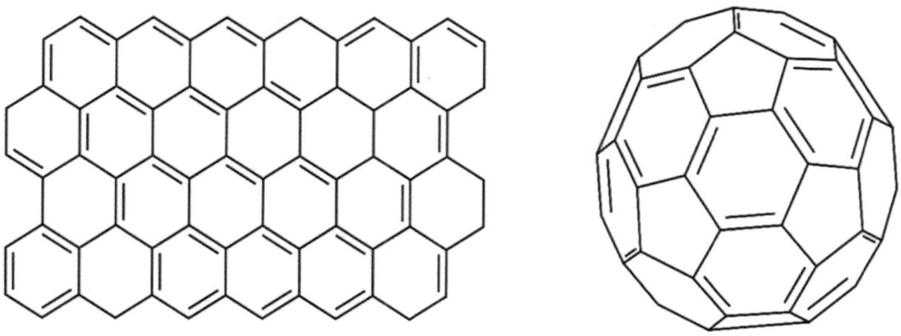

Figure 1.
General structure of the graphene and fullerene.

Fullerenes are spherical carbon nanomaterial derivatives. This structure, unlike other carbon structures, consists of a closed form of pentagonal and hexagonal carbon structures fused together [1].

Graphene nanomaterials are called two-dimensional nanomaterials because they consist of only one layer, while fullerenes are classified as zero dimensional closed cage type nanomaterial. The spherical form of the fullerene nanomaterial gives this nanomaterial a large surface area [8–10]. This feature is a sought after feature for biosensors and sensor systems. The most advantageous feature of the double bonds formed by the carbon structure is that they can be modified as they can easily respond to chemical reactions. Most of the chemical reactions occurred by the nucleophilic attacks can form active sides on the fullerene sphere. With these modifications, fullerenes can be chemically modified. Chemical modification is important for biomolecule immobilization or surface modification. Fullerene in 60 carbons has the capacity to form 30 bonds due to its spherical structure and 30 double bonds. These bonds can be modified with different chemical agents to form fullerene derivatives.

2.2 Fullerene derivatives and chemical reactions

The most isomerically found types of fullerene derivatives are C_{60} and C_{70} fullerenes. These derivatives can only be dissolved in highly non-polar liquids such as toluene and benzene. Thanks to this hydrophobic property, it can be used as a drug carrier, intercalator or modification material in hydrophobic layers as lipid layers. In computer studies about fullerene, solubility attempts that were made in 75 different solvents were examined [11].

Although it is possible to dissolve in different solvents, surface modification of fullerenes for biosensor and sensor technology is the most effective way to use fullerene. The solubility in water can be achieved by being modified with polar groups. The method developed by Hirsch and colleagues, fullerene was modified by 18 carboxyl groups, gained solubility in water as 34 mg/mL at pH = 7.4 [12]. In these studies, a nucleophilic cyclopropanation procedure was performed; the protection was removed with the help of bis-(polyamide)-malonate dendrimer; and 18 carboxyl groups were created (**Figure 2**). With this modification, a material, which can be used as an advantage in water solubility for modification of the carboxyl group, especially for sensor and biosensor technology, is obtained. With the activation of EDC/NHS, 18 carboxyl groups can be made to bind the amino group or a fullerene nanomaterial containing carboxyl groups can be modified on an amino group-modified transducer.

In another study, Cusan and his colleagues synthesized three ethylene glycol and three ammonium groups in poly charged fullerene-pyrrolidone derivatives [13]. In this study, which was carried out on two strategies, 2,2′-(ethylenedioxy)diethylamine was reacted by amino esterification with benzyl bromoacetate. Subsequently, the substance was interacted with carboxylate groups on fullerene to obtain fulleropyrrolidine in toluene. In this method, the authors have shown that purification is difficult and yield is low. For fullerene-PEG, in terms of biosensor and sensor technology, the fullerene derivative has two different arms with an amino group that can be used for modification by activating amino groups with glutaraldehyde a cross linker (**Figure 3**).

Fullerene modifications can be carried out entirely through the modification of the C = C bonds on fullerene. Fullerene modifications and derivatives of these modifications are seen in several studies. Accordingly, fullerene can be used to develop biosensors and sensors after being modified [14].

Apart from their use for modification material, another interesting feature of fullerene and fullerene-like materials is their photocatalytic advantage. The C_{60} shows a semiconductor-diode-like behavior and shows photo activity when

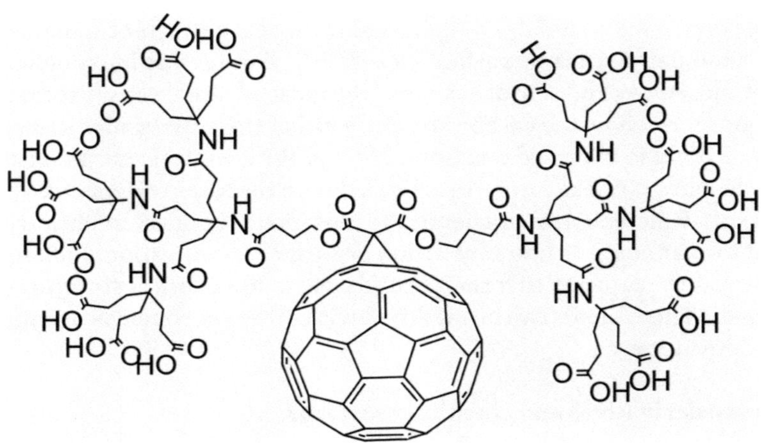

Figure 2.
Poly-carboxylated (18) fullerene.

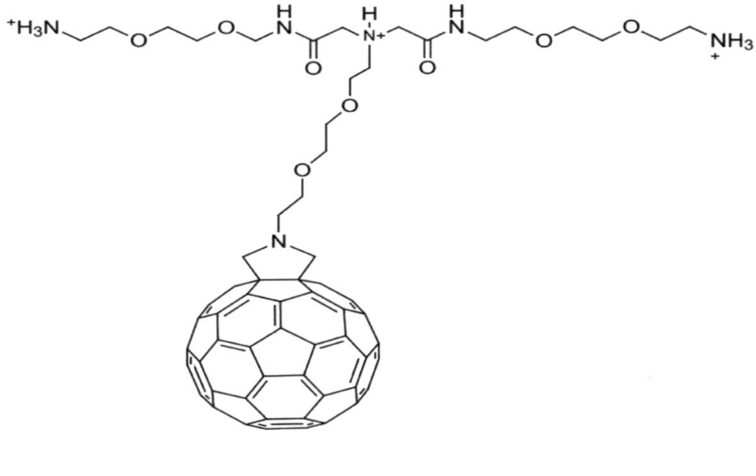

Figure 3.
Poly-charged fullerene derivative.

illuminated at a value close to 1.4 eV [1]. This photocatalytic feature shows that it can be used as a photo catalyst in optical sensors and biosensors.

In conclusion, due to its catalytic properties, the use of fullerene is not only used as an immobilization material but also it is used in biosensor and sensor systems.

3. Fullerene-modified sensors and biosensors

Sensor and biosensor technology is an important start line in the development of miniature analyzers. This technology is divided into two classes depending on the interaction of the sensor and analyte molecule: if the analyte molecule is transformed on the sensor surface called as catalytic based and if the analyte interacts with the surface called affinity-based sensors and biosensors [5]. The measurement can be performed electrochemically, optically, and piezoelectrically. In electrochemical sensors, electrodes are used as transducers. Different types of electrodes can be used according to the modification and measurement principle. The electrodes can be gold, carbon, platinum, and their derivatives. It is desired to create a modification layer over self-assembly monolayer; thiol containing chemicals can be used for gold and gold derivative electrodes [15]. Carbon derivative electrodes can be used for polymer production and adsorption type studies [16]. Thus, measurement is carried out with electrodes modified with the immobilized recognition agent. If the recognition agent is a catalytic agent (enzyme or nanoparticle), the electroactive species are released as a result of the target molecule that is transformed by recognition receptor. The method of measurement may also vary depending on the type of these species. For example, an amperometric technique is used if an electroactive species is formed, or a potentiometric measurement method is used if an ion is formed. In addition, interaction-based measurement is desired without a reaction between the analyte molecule and the recognition receptor on the electrode surface (DNA, antibody, MIP, polymer, etc.), and impedimetric techniques can be used [15, 17, 18].

Optically designed sensor and biosensor systems use optic systems with optic sensitivity as transducers or surface plasmon resonance systems, which are a highly sensitive system that is used quite frequently today with laser canteen systems. The photons detected by the transducer with chemical reaction light sensing capability that occurs in optic systems can be converted into meaningful signals. In cantilever systems, measurement is performed by creating differences in the angle of the reflected laser signal as a result of the analyte molecule, which is attached to the surface of the cantilever, whose laser signal is reflected under a lever, changes the oscillation of the lever [19]. In surface plasmon resonance type sensors, the laser signal reflected behind the gold bit surface changes with the binding of the target molecule to the dielectric constant of this gold surface [20]. Here, too, the main purpose is the interaction between the analyte molecule and the recognizing receptor. If there is a catalytic effect, a photon sensing transducer, if there is an affinity-based measurement, cantilever or SPR sensor systems are used.

Apart from these two techniques, piezoelectric systems, which are mass detection systems, can be designed as affinity-based instead of catalytic in biosensor and sensor systems. The method principle is the analyte molecule interacts with piezoelectric crystals used as transducers can generate signals by changing the oscillation of the piezoelectric systems [21].

In conclusion, the interaction between the target molecule and the recognition receptor on sensor and biosensor systems can be measured with electrical, optic, and piezoelectric systems. The important point in these measurements is the characteristic of the modification layer on which the recognizing receptor is immobilized. As we mentioned above, the usage of nanomaterials in biosensor and sensor systems is to increase the surface area and electrical conductivity, to create more stable layer for immobilization, and to use nanomaterials catalytically by generating electroactive signals from the electrical characteristics of nanomaterials. With the advantages of these materials, more sensitive and selective sensor and biosensor systems can be developed.

3.1 Fullerene-modified sensors

As we mentioned above, sensor systems can use molecularly imprinted polymers, nanocomposites, and other polymeric or dendrimer materials as recognition agent, instead of a biomolecule from biological source, and measurement basis can be formed with it. In this section, examples of this type of sensor technologies are given with the use or modifications of fullerene nanoparticles.

Zhong and colleagues performed the fullerene nanoparticles for catalytic activities, fullerenes were covalently modified with cysteine, and then palladium nanoparticles were added to this spherical structure to create Pd@Cys-C_{60} structure (**Figure 4**) to glucose detection without enzyme [22]. Firstly, the fullerene covalently bonded with the alpha amino group of cysteine on the nucleophilic attack basis in the medium containing the nanoparticles NaOH and EtOH. Then palladium cation was added, and Pd@Cys-C_{60} nanocomposite material was obtained. As is known, palladium is highly effective catalytic material in nano form. In this way, the researchers modified the glassy carbon electrode with this nanocomposite and developed a sensor system that can measure glucose between 2.5 μM and 1 mM, at the lowest detection limit of 1 μM. In selectivity studies, it gave 4.8% signal to substances such as fructose and ascorbic acid. As a result, sensitive and stable signals were obtained with fullerene, such as glucose sensors modified with previously made palladium nanoparticles [23].

Anusha et al. used fullerene nanomaterials with bimetallic nanoparticles to develop electrochemical sensors for vitamin D_3 determination [24]. Copper and nickel metal nanoparticles were used in this study. Glassy carbon electrode was used as the working electrode, and this electrode was modified by dropping C_{60} in toluene. The C_{60} was further reduced in the solution KOH solution and then to the solutions containing CuNPs, NiNPs, and nanoparticles were deposited electrochemically on this electrode, respective electrodeposition steps. Modifying the electrode with these nanoparticles is due to their surface area enhancing and catalytic effects. VitD_3 was measured by cyclic voltammetry by the electrochemical oxidation on NiNPs-CuNPs-C_{60}-GCE. With the development of the method, 1.25–475 μM

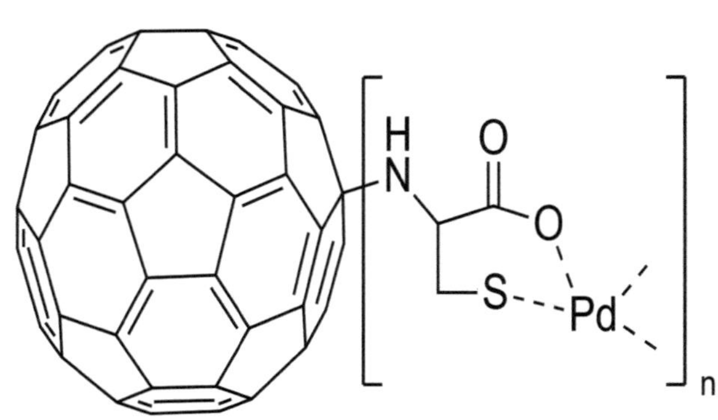

Figure 4.
Palladium nanoparticles and cysteine-modified fullerene particles (Pd@Cys-C60).

linearity and 0.0025 μM LOD values were achieved. As a result of the study, a more sensitive method was developed than similar methods [25–28].

Saha and Das have developed a moisture sensor with C_{60} nanoparticles that immobilized on thick alumina layer [29]. In this study, they created nano-sized cavities on the sensor surface with the use of fullerene nanoparticles. With these gaps, they have improved the surface area and have developed a sensor that can detect even trace amount of moisture not in literature by fullerene advantage. The sensor was briefly prepared by heating a ceramic layer at 900° after being modified with fullerene, alumina, and polyvinyl alcohol, respectively. The sensor enables capacitive measurements to detect moisture at the ppm levels, in the range of 1–25 ppm. Supported with fullerene has the potential to be used successfully in the gas, oil, and food drying industries, due to the small pore sizes, a highly selective sensor has been developed for other volatile liquids.

Shetti and colleagues have developed a fullerene-modified GCE biosensor for electrochemical acyclovir (ACV) determination [30]. The GCE modification with fullerene was carried out according to the drying process on the previously mentioned GCE method. In this study, fullerene was dissolved in dichloromethane instead of toluene, dried by dropping on GCE and reduced in KOH. Afterwards, the electrode was immersed in ACV and the accumulation of ACV on the electrode surface was carried out electrochemically. ACV measurement was carried out with DPV, and real sample experiments were carried out by adding ACV to the real samples as spike. With this sensor, ACV linear measurement was achieved between 90 nM and 6 μM and 1.48 nM lower limit. The sensor gave more sensitive results compared to other ACV measurement methods by modification with fullerene [31–35]. Tartaric acid made the most interference on the sensor with 11.6%. In real sample experiments, a matrix effect of less than 3% is observed.

Ertuğrul Uygun and colleagues have developed an impedimetric sensor system modified with cortisol-imprinted polymers on fullerene for the determination of cortisol in saliva [36]. In this study, fullerene C_{60} was dispersed in dimethylformamide and dried on a carbon screen printed electrode. COOH groups were then formed on fullerene by oxidation process in 2 M H_2SO_4. Afterwards, acrylamide was polymerized with APS around cortisol via these COOH groups, and cortisol-imprinted polymers were synthesized on the electrode surface. After removing cortisol with acetic acid, the CE-C_{60}-polyAcry sensor is ready for the determination of cortisol (**Figure 5**). In this study, which was performed in real saliva samples and compared with tandem mass spectrometry, the cortisol measurement was performed between 0.5 and 64 nM and the lowest detection limit of 0.14 nM was achieved. Modification of fullerene with carboxyl groups facilitated the synthesis

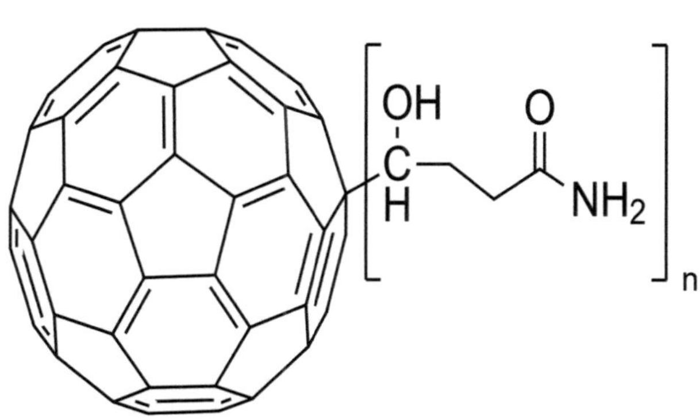

Figure 5.
C60-acry modified fullerene material.

of polymers and increased surface area. Sensor regression analysis compared to tandem mass spectrometry was found to be 0.9778.

Another sensor study developed the preparation of molecularly imprinted polymers by making them more advantageous with fullerene for the determination of ATP by Sharma and his colleagues [37]. In this study, fullerene C_{60} nanomaterials were primarily modified with amide derivatives. After this modification, fullerene derivatives assembled around ATP by crosslinking them with Pd nanoparticles (**Figure 6**). The fullerene derivatives used herein are amide, carboxyl, and uracil functionalized fullerene nanomaterials. This sensor is characterized by complex measuring systems. Concentration determination was carried out capacitive, and characterization of the polymer layer was performed piezoelectrically and voltammetrically. The ATP measurement was carried out piezoelectrically. 0.062 to 1.0 mM linear ATP was measured capacitively, and the lowest detection limit was 0.31 mM.

3.2 Fullerene-modified biosensors

Biosensor systems use biorecognition agents from biological origin as the recognizing agent. These systems can also be catalytic and affinity-based just as mentioned before. Enzymes are used in catalytic biosensors. The important point here is a catalytic biosensor system can measure the electroactive species or mediators released by enzymatic reaction. In affinity-based biosensors, the basis of the measurement is based on the measurement of the interaction between the biorecognition receptor and the target molecule. These interactions are DNA-DNA, DNA-protein, antigen-antibody, or protein-ligand. In this section, fullerene-modified biosensor systems are discussed.

Pan and Shih fullerenes have developed a piezoelectric-based immunosensor system with C60 nanomaterials [38]. In this system, piezoelectric quartz crystals were modified by spin-coating the toluene solution containing polyvinyl chloride and fullerene onto these crystals. An adsorption type modification was carried out by adding anti-immunoglobulin G on this modification. In this biosensor, IgG determination reached the determination range between 0.0001 and 0.01 mg/mL

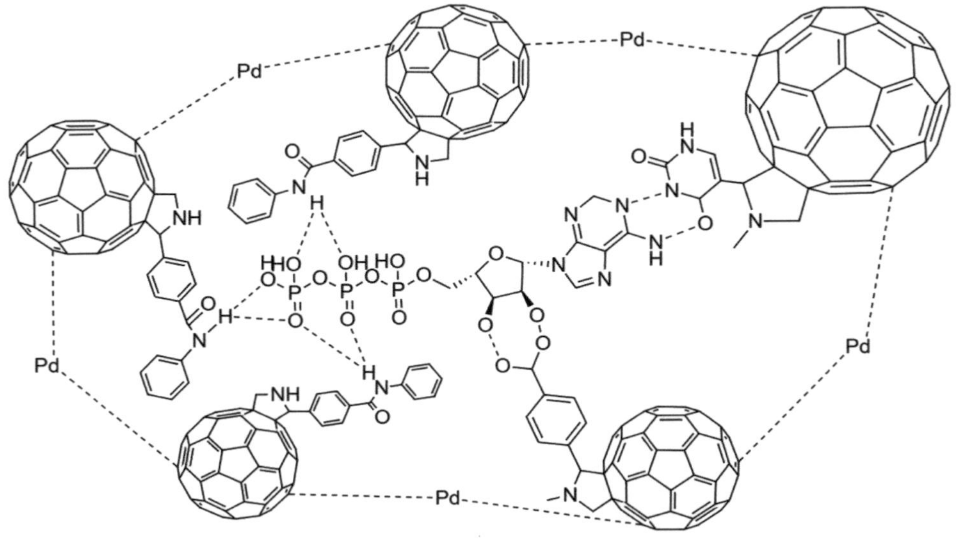

Figure 6.
Amide, carboxyl, and uracil modified fullerene polymers for ATP imprinted sensor.

and the lowest detection limit below 0.0001 mg/mL and has shown a very high selectivity.

Suresh and colleagues have developed another C_{60} modified immunosensor [39]. With this sensor, prostate specific antigen was determined, and the lower measurement limit of 0.002 ng/mL was reached. The biosensor can detect PSA in a measuring range from 0.005 to 20 ng/mL. GCE was used as the working electrode in the study. The C_{60} dissolved in toluene and dried on GCE. CuNPs were then deposited electrochemically on fullerene which was electrochemically reduced in NaOH. Then, the electrode was immersed in hydroquinone (HQ) solution and electrochemically bound to surface by the application of potential. After that, HQ-modified electrode was activated with EDC/NHS, and anti-PSA was covalently immobilized to this surface. PSA was determined by measuring the hydrogen peroxide reduction of HRP by forming a layer-by-layer on electrode by blocking the active ends exposed with BSA and the secondary antibody labeled with PSA and HRP, respectively. Electrochemical characterization was performed by CV and EIS. The measurement was carried out with the CV method by the principle of measuring the reduction of the HQ. The biosensor in serum showed a more sensitive result than similar studies with its fullerene modification, showing a 2% matrix effect [40–45].

Zhou et al. performed an ultra-trace amount of miRNA-141 measurement in another fullerene-modified biosensor study [46]. In this study, a new method has been developed that provides a double signal increase. G-quadruplex, which is complementary with miRNA-141, was combined to form the DNA-RNA hybrid. The DNA fragment was then cut with a duplex-specific nuclease, and miRNA-141 was released to triggering the next step and measuring. The reason of the system's ultra-sensitivity is that by C_{60}-modified gold electrodes modified with amino and thiol groups. The fullerene dissolved in toluene, then passed to the water phase by removing toluene. PTC-NH_2 was obtained by adding anhydrous ethylenediamine to the solution containing 3,4,9,10-perylenetetracarboxylic dianhydride (PTCDA) (**Figure 7**). PTC-NH_2 was mixed with the fullerene in the aqueous phase to form amid groups on the fullerene. Afterwards, more active groups were formed by adding EDC/NHS on C_{60} by activated Cys. Gold electrode modification

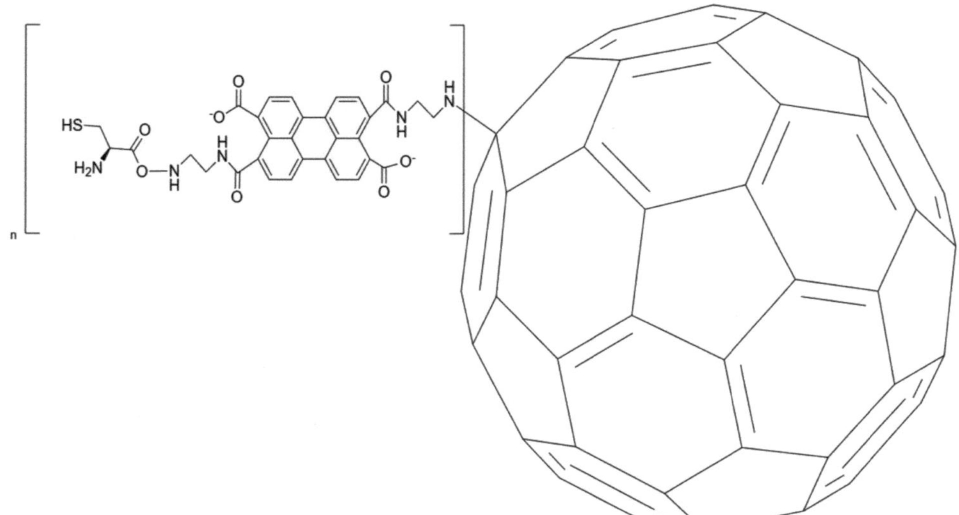

Figure 7.
SH and NH_2 modified PTC-NH_2-C_{60}.

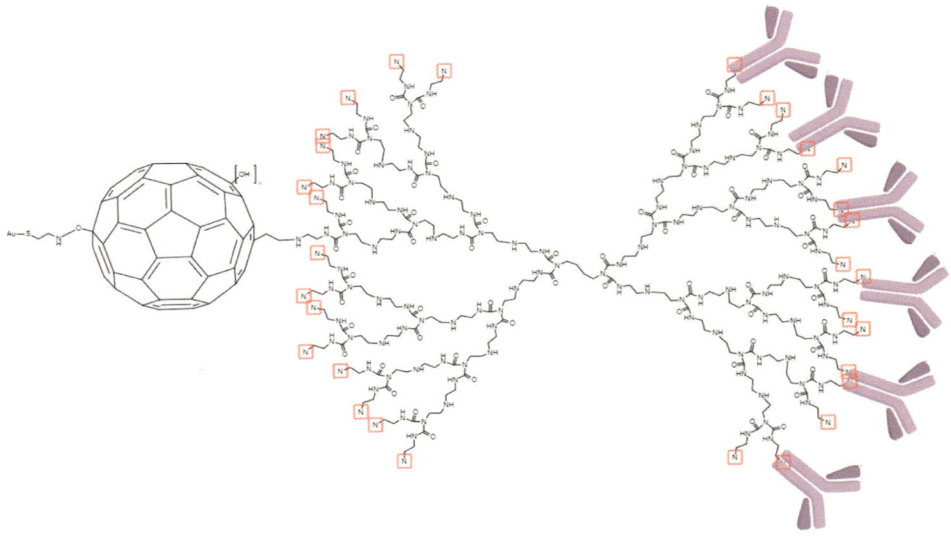

Figure 8.
Modification layers of the fullerene-PAMAM modified anti-Fetuin-A biosensor (red indicated nitrogen groups represent the active immobilization residue).

Analytical method	Modification	Analyte	Linear detection range	LOD	Reference
Chronoamperometry	Pd@Cys-C_{60}	Glucose	2.5 μM–1 mM	1 μM	[22]
Cyclic voltammetry	NiNPs-CuNPs-C_{60}	Vitamin D_3	1.25–475 μM	0.0024 μM	[24]
Capacitive	C_{60}-Alumina	Moisture	1–25 ppm	1 ppm	[29]
Differential pulse voltammetry	ACV-C_{60}	Acyclovir	90 nM–6 μ M	1.48 nM	[30]
Electrochemical impedance spectroscopy	C_{60}-poly-acrylamide	Cortisol	0.5–64 nM	0.14 nM	[36]
Capacitive	Amide-C_{60}, carboxyl-C_{60}, uracil-C_{60}	ATP	0.062–1 mM	0.31 mM	[37]
Piezoelectric	Anti-IgG-PVA-C_{60}	IgG	0.0001–0.01 mg/mL	0.0001 mg/mL	[38]
Cyclic voltammetry	C_{60}-CuNPs-HQ-anti-PSA	PSA	0.005–20 ng/mL	0.002 ng/mL	[39]
Differential pulse voltammetry	C_{60}-PTC-Cys	miRNA-141	0.1 pM–100 nM	7.78 fM	[46]
Electrochemical impedance spectroscopy	4ATP- C_{60}-polyOH-PAMAM-anti-FetuinA	Fetuin A	5–400 ng/mL	1.44 ng/mL	[15]
Piezoelectric	C_{60}-GOx	Glucose	100 μM–10 mM	39 μM	[47]

Table 1.
Comparison of the fullerene-modified sensor and biosensor technology.

was carried out by forming the Au-SH bond of SH groups on the modified C_{60}. Electrochemical characterization of the biosensor was carried out by EIS and CV, and differential pulse voltammetry (DPV) was used as the method of analysis. The target miRNA-141 was determined with this biosensor to measure between 0.1 pM and 100 nM, and the lowest detection limit of 7.78 fM was achieved. As a result of increasing the surface area with fullerene on developed biosensor, the researchers have reached a very low detection limit.

Uygun et al. have developed an impedimetric Fetuin-A biosensor, modified with Fullerene-PAMAM (G5) [15]. In this study, the gold electrode, as a working electrode, was first modified by forming self-assembly mono layer of 4-aminothiophenol (4-ATP). Then, activated by EDC/NHS poly-hydroxylated fullerene was firstly dropped to this surface. Au-4ATP-C_{60}-OH was modified by PAMAM (G5) dropped to the surface and anti-Fetuin-A antibodies were dropped to the surface, respectively (**Figure 8**). The biosensor compared to ELISA provided linear measurement between 5 and 400 ng/mL and the lowest 1.44 ng/mL measurement opportunity. The fullerene-PAMAM modification has shown that the biosensor is more advantageous than ELISA due to its three-minute determination and high surface area.

Chuang et al. measured glucose with the piezoelectric system using C_{60} modified with glucose oxidase enzyme [47]. After the enzyme fullerene and glucose oxidase was incubated for 70 hours, it was immobilized on the piezoelectric crystal. In the sensor system where glucose was measured, LOD was found to be 39 μM. It is stated that in the system, which has a linear glucose measurement between 100 μM and 10 mM, glucose measurement can be made in real samples.

Sensor and biosensor studies have shown that the use of fullerene has been an advantage in these systems. It is summarized in **Table 1** with the complete example.

4. Conclusions

Nanomaterials are becoming more important and interesting day by day and their usage areas are increasing. In this book section where the sensor and biosensor systems modified with fullerene are explained, how the fuller is adapted to these systems and how it is modified is explained with examples. As a result, fullerene has successfully completed the task and contributed significantly to the development of these systems. The use of fullerene enabled the development of sensitive sensors by increasing the surface area. At the same time, its contribution to being an indestructible immobilization material and providing durability cannot be ignored.

Acknowledgements

This chapter was dedicated to our lovely son Bora Metin UYGUN. All figures were created by using https://chemdrawdirect.perkinelmer.cloud/rest/online program.

Conflict of interest

Authors have no conflict of interest.

Acronyms and abbreviations

4-ATP	4-aminothiophenol
APS	ammonium persulfate
Au	gold

BSA	bovine serum albumin
C60	fullerene
COOH	carboxyl
Cu	cupper
CuNPs	copper nanoparticles
CV	cyclic voltammetry
Cys	cysteine, cysteamine
DPV	differential pulse voltammetry
EDC	N-(3-dimethylaminopropyl)-N′-ethylcarbodiimide
EIS	electrochemical impedance spectroscopy
ELISA	enzyme-linked immunosorbent assay
EtOH	ethanol
GCE	glassy carbon electrode
GOx	glucose oxidase
HQ	hydroquinone
HRP	Horse radish peroxidase
KOH	potassium hydroxide
miRNA	micro RNA
mL	milliliter
NaOH	sodium hydroxide
ng	nanogram
NHS	N-hydroxysuccinimide
Ni	nickel
NPs	nanoparticles
PAMAM	poliamidoamin
PSA	prostate specific antigen
PTCDA	3,4,9,10-perylenetetracarboxylic dianhydride
QCM	quartz crystal microbalance
SH	thiol

Author details

Hilmiye Deniz Ertuğrul Uygun[1] and Zihni Onur Uygun[2]*

1 Nanoscience and Nanoengineering Department, Dokuz Eylül University, İzmir, Turkey

2 Medical Biochemistry Department, Faculty of Medicine, Ege University, İzmir, Turkey

*Address all correspondence to: onur.uygun@ege.edu.tr; onur_uygun@hotmail.com

References

[1] Baena JR, Gallego M, Valcárcel M. Fullerenes in the analytical sciences. TrAC, Trends in Analytical Chemistry. 2002;**21**:187-198. DOI: 10.1016/S0165-9936(02)00306-0

[2] Lim E-K, Kim T, Paik S, Haam S, Huh Y-M, Lee K. Nanomaterials for theranostics: Recent advances and future challenges. Chemical Reviews. 2015;**115**:327-394. DOI: 10.1021/cr300213b

[3] Masatoshi HS, Daisuke I. Applications of Nanomaterials in Sensors and Diagnostics2013. DOI: 10.1007/978-3-642-36025-1

[4] Rinken T. State of the art in biosensors - general aspects. Libro. 2013:360. DOI: 10.5772/45832

[5] Turner APF. Biosensors: Fundamentals and Applications. Oxford University Press; 1987

[6] Ye L, Mosbach K. Molecularly imprinted microspheres as antibody binding mimics. Reactive and Functional Polymers. 2001;**48**:149-157. DOI: 10.1016/S1381-5148(01)00050-5

[7] Sotiropoulou S, Gavalas V, Vamvakaki V, Chaniotakis N. Novel carbon materials in biosensor systems. Biosensors & Bioelectronics. 2003;**18**:211-215. DOI: 10.1016/S0956-5663(02)00183-5

[8] Luo B, Liu G, Wang L. Recent advances in 2D materials for photocatalysis. Nanoscale. 2016;**8**:6904-6920. DOI: 10.1039/c6nr00546b

[9] Novoselov KS, Geim AK, Morozov SV, Jiang D, Zhang Y, Dubonos SV, et al. Electric field in atomically thin carbon films. Science. 2004;**306**:666-669. DOI: 10.1126/science.1102896

[10] Chen M, Guan R, Yang S. Hybrids of Fullerenes and 2D Nanomaterials2019. p. 1800941. DOI: 10.1002/advs.201800941

[11] Toropov AA, Leszczynska D, Leszczynski J. QSPR study on solubility of fullerene C60 in organic solvents using optimal descriptors calculated with SMILES. Chemical Physics Letters. 2007;**441**:119-122. DOI: 10.1016/j.cplett.2007.04.094

[12] Brettreich M, Hirsch A. A highly water-soluble dendro[60]fullerene. Tetrahedron Letters. 1998;**39**:2731-2734. DOI: 10.1016/S0040-4039(98)00491-2

[13] Cusan C, Ros D, Spalluto G, Foley S, Janot J, Seta P, et al. FULL PAPER a New Multi-Charged C 60 Derivative: Synthesis and Biological Properties; 2002. pp. 0-6

[14] Hermanson GT. Bioconjugate Techniques. Elsevier Inc.; 2008. DOI: 10.1016/B978-0-12-370501-3.X0001-X

[15] Uygun ZO, Şahin Ç, Yılmaz M, Akçay Y, Akdemir A, Sağın F. Fullerene-PAMAM(G5) composite modified impedimetric biosensor to detect Fetuin-A in real blood samples. Analytical Biochemistry. 2018;**542**:11-15. DOI: 10.1016/J.AB.2017.11.007

[16] Uygun ZO, Dilgin Y. A novel impedimetric sensor based on molecularly imprinted polypyrrole modified pencil graphite electrode for trace level determination of chlorpyrifos. Sensors and Actuators B: Chemical. 2013;**188**:78-84. DOI: 10.1016/j.snb.2013.06.075

[17] Uygun HD, Onur Z, Uygun E. A novel chronoimpedimetric glucose sensor in real blood samples modified by glucose-imprinted pyrrole-aminophenylboronic acid modified

screen printed. Electroanalysis. 2019:1-5. DOI: 10.1002/elan.201900537

[18] Ertuğrul HD, Uygun ZO. Impedimetric Biosensors for Label-Free and Enzymless Detection. Interferon; 2016. p. 13. DOI: 10.5772/57353

[19] Li C, Ma X, Guan Y, Tang J, Zhang B. Microcantilever array biosensor for simultaneous detection of carcinoembryonic antigens and a-fetoprotein based on real-time monitoring of the profile of cantilever. ACS Sensors. DOI: 10.1021/acssensors.9b01604

[20] Kumar A, Kim S, Nam JM. Plasmonically engineered nanoprobes for biomedical applications. Journal of the American Chemical Society. 2016;**138**:14509-14525. DOI: 10.1021/jacs.6b09451

[21] Ávila M, Zougagh M, Ríos Á, Escarpa A. Molecularly imprinted polymers for selective piezoelectric sensing of small molecules. TrAC, Trends in Analytical Chemistry. 2008;**27**:54-65. DOI: 10.1016/j.trac.2007.10.009

[22] Zhong X, Yuan R, Chai Y. In situ spontaneous reduction synthesis of spherical Pd@Cys-C 60 nanoparticles and its application in nonenzymatic glucose biosensors. Chemical Communications. 2012;**48**:597-599. DOI: 10.1039/c1cc16081h

[23] Lu LM, Li HB, Qu F, Zhang XB, Shen GL, Yu RQ. In situ synthesis of palladium nanoparticle-graphene nanohybrids and their application in nonenzymatic glucose biosensors. Biosensors & Bioelectronics. 2011;**26**:3500-3504. DOI: 10.1016/j.bios.2011.01.033

[24] Anusha T, Bhavani KS, Kumar JVS, Brahman PK. Designing and fabrication of electrochemical nanosensor employing fullerene-C60 and bimetallic nanoparticles composite film for the detection of vitamin D3 in blood samples. Diamond and Related Materials. 2020;**104**:107761. DOI: 10.1016/J.DIAMOND.2020.107761

[25] Canevari TC, Cincotto FH, Landers R, Machado SAS. Synthesis and characterization of α-nickel (II) hydroxide particles on organic-inorganic matrix and its application in a sensitive electrochemical sensor for vitamin D determination. Electrochimica Acta. 2014;**147**:688-695. DOI: 10.1016/j.electacta.2014.10.012

[26] Abbar JC, Nandibewoor ST. Development of electrochemical method for the determination of chlorzoxazone drug and its analytical applications to pharmaceutical dosage form and human biological fluids. Industrial and Engineering Chemistry Research. 2012;**51**:111-118. DOI: 10.1021/ie2021812

[27] Nallbani A, Holubová J, Sýs M, Arbneshi T, Vytřas K. Voltammetric determination of cholecalciferol at glassy carbon electrode performed in water-ethanol mixture. Potravinarstvo Slovak Journal of Food Sciences. 2018;**12**:166-172. DOI: 10.5219/889

[28] Men K, Chen Y, Liu J, Wei D. Electrochemical detection of vitamin D2 and D3 based on a Au-Pd modified glassy carbon electrode. International Journal of Electrochemical Science. 2017;**12**:9555-9564. DOI: 10.20964/2017.10.15

[29] Saha D, Das S. Development of fullerene modified metal oxide thick films for moisture sensing application. Materials Today: Proceedings. 2018;**5**:9817-9825. DOI: 10.1016/J.MATPR.2017.10.172

[30] Shetti NP, Malode SJ, Nandibewoor ST. Electrochemical behavior of an antiviral drug acyclovir

at fullerene-C60-modified glassy carbon electrode. Bioelectrochemistry. 2012;**88**:76-83. DOI: 10.1016/J. BIOELECHEM.2012.06.004

[31] Jankowski A, Jankowska AL, Lamparczyk H. Determination and pharmacokinetics of acyclovir after ingestion of suspension form. Journal of Pharmaceutical and Biomedical Analysis. 1998;**18**:249-254. DOI: 10.1016/S0731-7085(98)00164-2

[32] Peh KK, Yuen KH. Simple high-performance liquid chromatographic method for the determination of acyclovir in human plasma using fluorescence detection. Journal of Chromatography. B, Biomedical Sciences and Applications. 1997;**693**:241-244. DOI: 10.1016/s0378-4347(97)00041-8

[33] Heli H, Zarghan M, Jabbari A, Parsaei A, Moosavi-Movahedi AA. Electrocatalytic oxidation of the antiviral drug acyclovir on a copper nanoparticles-modified carbon paste electrode. Journal of Solid State Electrochemistry. 2010;**14**:787-795. DOI: 10.1007/s10008-009-0846-x

[34] Wang F, Chen L, Chen X, Hu S. Studies on electrochemical behaviors of acyclovir and its voltammetric determination with nano-structured film electrode. Analytica Chimica Acta. 2006;**576**:17-22. DOI: 10.1016/j. aca.2005.12.023

[35] Swart KJ, Hundt HKL, Groenewald AM. Automated high-performance liquid chromatographic method for the determination of acyclovir in plasma. Journal of Chromatography. A. 1994;**663**:65-69. DOI: 10.1016/0021-9673(94)80496-6

[36] Ertuğrul Uygun HD, Uygun ZO, Canbay E, Sağın FG, Sezer E. Non-invasive cortisol detection in saliva by using molecularly cortisol imprinted fullerene-acrylamide modified screen printed electrodes. Talanta. 2020;**206**:120225. DOI: 10.1016/j. talanta.2019.120225

[37] Sharma PS, Dabrowski M, Noworyta K, Huynh T-P, KC CB, Sobczak JW, et al. Fullerene derived molecularly imprinted polymer for chemosensing of adenosine-5′-triphosphate (ATP). Analytica Chimica Acta. 2014;**844**:61-69. DOI: 10.1016/J. ACA.2014.07.005

[38] Pan N-Y, Shih J-S. Piezoelectric crystal immunosensors based on immobilized fullerene C60-antibodies. Sensors and Actuators B: Chemical. 2004;**98**:180-187. DOI: 10.1016/J. SNB.2003.09.034

[39] Suresh L, Bondili JS, Brahman PK. Development of proof of concept for prostate cancer detection: An electrochemical immunosensor based on fullerene-C60 and copper nanoparticles composite film as diagnostic tool. Materials Today Chemistry. 2020;**16**:100257. DOI: 10.1016/J.MTCHEM.2020.100257

[40] Kim DJ, Lee NE, Park JS, Park IJ, Kim JG, Cho HJ. Organic electrochemical transistor based immunosensor for prostate specific antigen (PSA) detection using gold nanoparticles for signal amplification. Biosensors & Bioelectronics. 2010;**25**:2477-2482. DOI: 10.1016/j.bios.2010.04.013

[41] Salimi A, Kavosi B, Fathi F, Hallaj R. Highly sensitive immunosensing of prostate-specific antigen based on ionic liquid-carbon nanotubes modified electrode: Application as cancer biomarker for prostate biopsies. Biosensors & Bioelectronics. 2013;**42**:439-446. DOI: 10.1016/j. bios.2012.10.053

[42] Chen Y, Yuan PX, Wang AJ, Luo X, Xue Y, Zhang L, et al. A novel electrochemical immunosensor

for highly sensitive detection of prostate-specific antigen using 3D open-structured PtCu nanoframes for signal amplification. Biosensors & Bioelectronics. 2019;**126**:187-192. DOI: 10.1016/j.bios.2018.10.057

[43] Ji S, Lee M, Kim D. Detection of early stage prostate cancer by using a simple carbon nanotube@paper biosensor. Biosensors & Bioelectronics. 2018;**102**:345-350. DOI: 10.1016/j.bios.2017.11.035

[44] Wei Q, Xiang Z, He J, Wang G, Li H, Qian Z, et al. Dumbbell-like Au-Fe3O4 nanoparticles as label for the preparation of electrochemical immunosensors. Biosensors & Bioelectronics. 2010;**26**:627-631. DOI: 10.1016/j.bios.2010.07.012

[45] Yang M, Javadi A, Gong S. Sensitive electrochemical immunosensor for the detection of cancer biomarker using quantum dot functionalized graphene sheets as labels. Sensors and Actuators B. 2011;**155**:357-360. DOI: 10.1016/j.snb.2010.11.055

[46] Zhou L, Wang T, Bai Y, Li Y, Qiu J, Yu W, et al. Dual-amplified strategy for ultrasensitive electrochemical biosensor based on click chemistry-mediated enzyme-assisted target recycling and functionalized fullerene nanoparticles in the detection of microRNA-141. Biosensors & Bioelectronics. 2020;**150**:111964. DOI: 10.1016/J.BIOS.2019.111964

[47] Chuang C-W, Shih J-S. Preparation and application of immobilized C60-glucose oxidase enzyme in fullerene C60-coated piezoelectric quartz crystal glucose sensor. Sensors and Actuators B: Chemical. 2001;**81**:1-8. DOI: 10.1016/S0925-4005(01)00914-5

Chapter 5

Cellulose Nanocrystals: From Classical Hydrolysis to the Use of Deep Eutectic Solvents

Manon Le Gars, Loreleï Douard, Naceur Belgacem and Julien Bras

Abstract

During the last two decades, interest in cellulosic nanomaterials has greatly increased. Among these nanocelluloses, cellulose nanocrystals (CNC) exhibit outstanding properties. Indeed, besides their high crystallinity, cellulose nanocrystals are interesting in terms of morphology with high aspect ratio (length 100–1000 nm, width 2–15 nm), high specific area, and high mechanical properties. Moreover, they can be used as rheological modifier, emulsifier, or for barrier properties, and their surface chemistry opens the door to numerous feasible chemical modifications, leading to a large panel of applications in medical, electronic, composites, or packaging, for example. Traditionally, their extraction is performed via monitored sulfuric acid hydrolysis, leading to well-dispersed aqueous CNC suspensions; these last bearing negative charges (half-sulfate ester groups) at their surface. More recently, natural chemicals called deep eutectic solvents (DESs) have been used for the production of CNC in a way of green chemistry, and characterization of recovered CNC is encouraging.

Keywords: cellulose, sulfuric acid hydrolysis, cellulose nanocrystals, bio-based nanomaterials, deep eutectic solvents

1. Introduction

As renewable and biodegradable nanomaterials, nanocelluloses have raised a huge interest for several decades. Indeed, their natural available and abundant source—the biomass—as well their interesting properties makes them materials of choice for new nanomaterial science in a large panel of applications. Two types of nanocelluloses exist: cellulose nanofibrils (CNF) and cellulose nanocrystals (CNC) differing from each other in their properties as well in their isolation process. As presented in **Figure 1**, the exponential evolution of the number of publications and patents dealing with cellulose nanocrystals confirms the large interest generated by these nanomaterials.

This chapter aims to describe CNC isolation from cellulose using classical or more recent methods, as well as their properties and applications.

IntechOpen

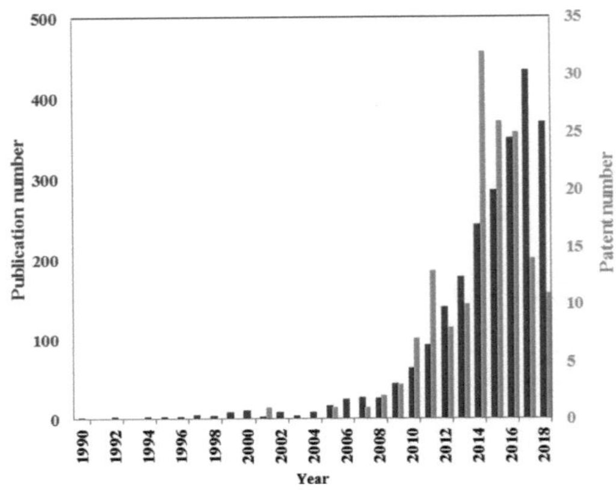

Figure 1.
Noncumulative evolution of the number of publications and patents dealing with CNC (source: SciFinder, April 2019—Descriptors, cellulose nanocrystal, cellulose nanorod, rodlike cellulose, cellulose nanowire, cellulose crystallite, cellulose nanoparticle, cellulose whiskers, nanocrystalline cellulose—Language, English).

2. Cellulose nanocrystals from classical acid hydrolysis

2.1 From cellulose to cellulose nanocrystals

2.1.1 Structure of cellulose

Cellulose can be extracted from a large variety of sources, kike wood (the main source), seed fibers (cotton), bast fibers (flax, jute, ramie), some animal species (tunicates), fungi, and fruits, with different cellulose contents [1]. With more than 10^{11} tons of cellulose produced each year [2], with less than 5% is extracted for applications, cellulose is the most abundant polymer on our planet [3]. Historically, cellulose was discovered after being extracted with nitric acid from several plants by the French researcher Anselme Payen in 1838, who characterized the residual compound with chemical formula $C_6H_{10}O_5$ [3, 4]. In 1939, the name "cellulose" was for the first time introduced in the scientific community. After almost 200 years of cellulose extraction, modification, and industrial use, this sustainable and biodegradable polymer is currently used for several applications, from paper and cardboard to biomedical, building, textile, cosmetics, pharmacy, and composites [5, 6]. Indeed, intrinsic properties of cellulose fibers—abundancy, renewability, and availability—as well its fibrillary structure or mechanical properties (strength, flexibility) make them materials of choice for such applications. Indeed, in their natural form, cellulose fibers are included in hemicellulose- and lignin-based matrix like a natural composite and acts as the primary compound of plant cell walls by providing high mechanical properties and maintaining their structure.

Looking more precisely at cellulose structure, it is a linear homopolymer of β-D-glucopyranose ($C_6H_{12}O_6$) units. These anhydro-D-glucose units (AGU) are linked by β-(1–4)-glycosidic linkage, a covalent bonding between equatorial OH group in C4 and C1 carbon atom of the next unit. Every unit is twisted at 180°C with respect to its surrounding environment and is in chair conformation, with the three hydroxyl groups in equatorial position. Cellobiose ($C_{12}H_{22}O_{11}$)—the combination of two anhydroglucose units (AGU)—is the repeating unit of cellulose [2, 3]. Cellulose general formula is represented in **Figure 2(a)**.

End groups of cellulose polymer are chemically different: one nonreducing end and one reducing bearing aldehyde group. Note that cellulose degree of

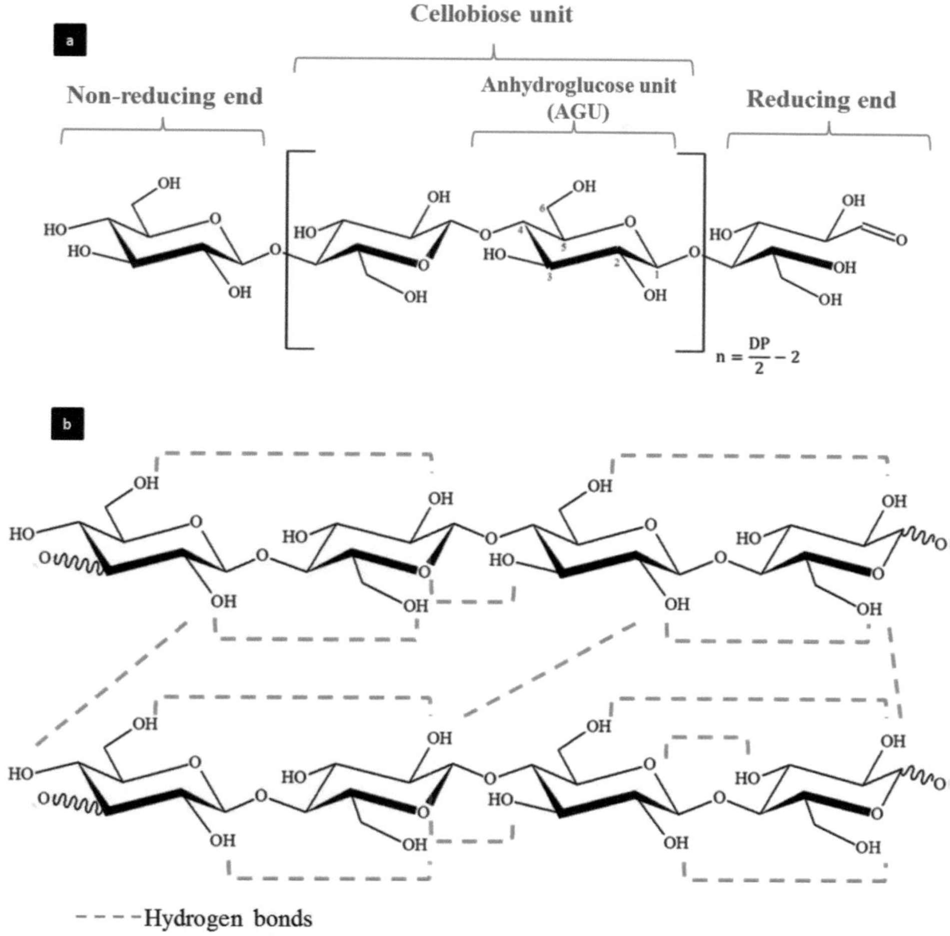

Figure 2.
(a) Chemical structure of cellulose chain and (b) representation of some hydrogen bonds between two cellulose chains.

polymerization (DP) is expressed as a function of the AGU unit number and depends on the cellulose source and isolation process (e.g., DP between 300 and 1700 for wood pulp and 800–1000 for cotton) [3]. The numerous hydroxyl groups—three per AGU—induce possible functionalization of cellulose as well as intra- and intermolecular hydrogen bonds in and between cellulose chains. These interactions form stabilized and flexible cellulose filaments: the cellulose microfibrils (see **Figure 3(a)**).

Moreover, cellulosic chains are rearranged into different regions: the ordered crystalline and the disordered amorphous ones. Indeed, a cellulose chain can be represented as a crystalline wire connected by amorphous areas (see **Figure 3(b)**). It explains the aggregation of cellulose chains and thus their arrangement into microfibrils. The latter are assembled in bundles, themselves assembled in cellulose fibers, with a semi-crystalline structure. Cellulose crystals present four polymorphs: cellulose I, II, III, and IV. Cellulose I is the most abundant form in nature and is present under cellulose Iα and Iβ forms whose ratio depends on the source and affect cellulose properties [3, 4]. Crystallinity of cellulose varies according to the source and is in the large range of 40–80% [3], leading to highly cohesive fibers. Looking more precisely at the level of the microfibrils, they are composed of elementary fibrils, with a diameter around 5 nm. This general structure provides

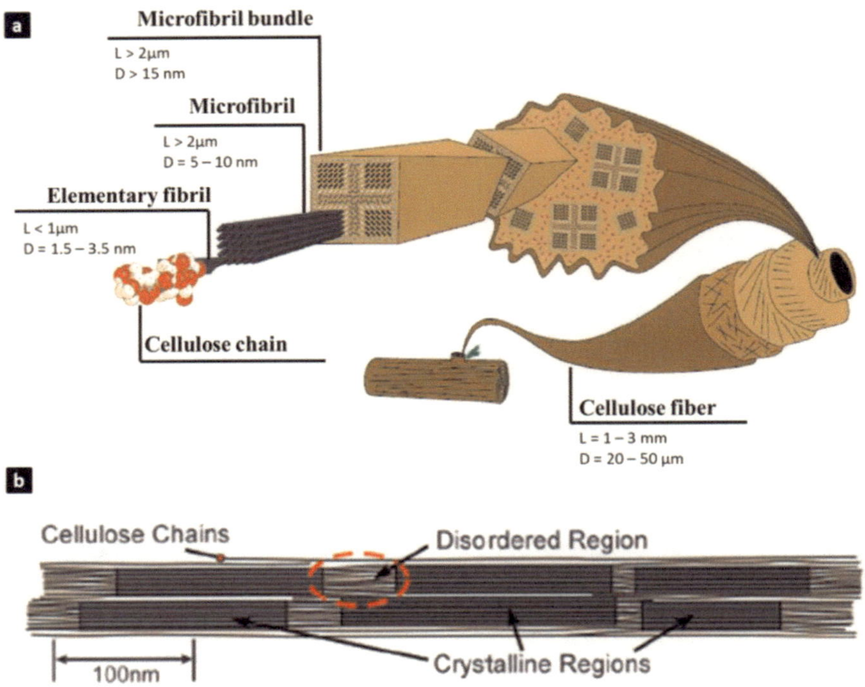

Figure 3.
Schematization of a simplified (a) composition of cellulose fiber (extracted and adapted from [7]) and (b) arrangements of crystalline and amorphous domains in cellulose chains (extracted from [8]).

visualization of the different scales inside the cellulose fiber. In addition to being environmentally relevant, cellulose fibers present interesting mechanical properties, ability for further surface modification, low toxicity, low cost, and other properties making them outstanding materials for a lot of traditional as well as innovative applications.

2.1.2 Cellulose nanocrystals from cellulose

Hierarchical structure of cellulose fibers is at the origin of cellulosic nanomaterials, the nanocelluloses, having at least one dimension of nanometric scale as their name suggests. Indeed, from elementary fibrils, two types of nanocellulose can be obtained differing by their isolation procedure, as well by their properties and applications. **Figure 4** shows their different morphologies. Briefly, cellulose nanofibrils are obtained by applying high shear mechanical treatment to a cellulose suspension, and recovered nanofibrils have a length between 500 nm and 10 μm and a width between 5 and 50 nm, according to their preparation method, their source, and potential chemical treatment.

First investigation on cellulose nanocrystal (CNC) was reported by Ranby et al. in 1950 [9, 10]: after carrying out a sulfuric acid hydrolysis to wood cellulose fibers, he observed rodlike particles with two nanoscale dimensions. Indeed, by hydrolyzing cellulose fibers, most of the amorphous parts of the cellulose are disintegrated, and final nanomaterials are highly crystalline. Cellulose nanocrystals have a length between 100 and 500 nm and a width between 2 and 15 nm depending on the cellulose source and the chemical treatment applied. Indeed, even if the use of sulfuric acid for hydrolysis is the most common process, other research groups have investigated the use of other acids, leading to CNC with different properties. In any case, washing steps are essential to remove any chemicals and to well-disperse the

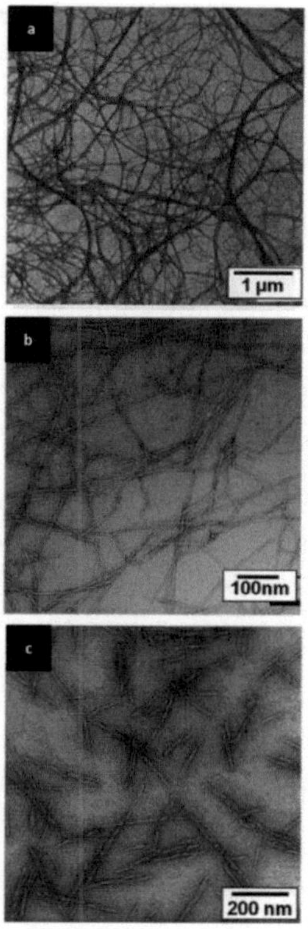

Figure 4.
TEM images of (a) microfibrillated cellulose (MFC), (b) TEMPO-oxidized nanofibrillated cellulose (NFC), and (c) wood cellulose nanocrystals (CNC) (extracted from [8]).

CNC. Concerning their industrialization, around 10 CNC producers can be recorded, with annual production up to 400 tons/year. These productions are significantly lower than those of CNF, but requirement of more chemicals and difficult industrial production steps (washing, dialysis, and sonication) can easily explain this difference. **Table 1** shows the non-exhaustive list of CNC producers

Company	Country	Annual production capacity (tons)
CelluForce Inc.	USA	400
America Process Inc. (GranBio)	USA	200
Alberta Pacific Forest Industries Inc.	Canada	180 (expected in 2021)
Anomera Inc.	Canada	11
Forest Products Laboratory (FPL)	USA	5
University of Maine	USA	4
Blue Goose Biorefineries Inc.	Canada	4
Cellulose Lab	Canada	4
Advanced Cellulosic Material Inc.	Canada	1
FPInnovations	Canada	0.5
InnoTech Alberta	Canada	0.3
Embrapa/National Nanotechnology Laboratory for Agriculture	Brazil	Pilot
Melodea	Israel	Pilot

Table 1.
Main CNC producers and their location and annual production capacity (data extracted from Refs. [12, 13]).

and their annual production capacity. Note that the leader and pioneer CelluForce© has recently announced new strategy of efficient industrialization [11].

2.2 Cellulose nanocrystals: nanomaterials with interesting properties

2.2.1 Isolation of cellulose nanocrystals

Cellulose nanocrystals are obtained by applying a chemical treatment to cellulose fibers: mild acid hydrolysis. Typically employing strong sulfuric acid H_2SO_4 is going to penetrate into accessible amorphous domains of cellulosic chains and dissolve them to release crystalline parts. Amorphous domains are randomly oriented and arranged inducing a lower density of these domains which are thus more vulnerable to acid hydrolysis [14] and especially to the infiltration of hydronium ions H_3O^+ leading to hydrolytic cleavage of glycosidic bonds. In this sense, Ranby et al. were the first to prove the preparation and the presence of CNCs, the smallest cellulosic building blocks. **Figure** 5 synthesizes the different steps of CNC isolation using sulfuric acid hydrolysis.

As previously mentioned, cellulose fibrils are exposed to a sulfuric acid hydrolysis, with defined concentration, temperature, and reaction time. Once amorphous domains are dissolved, a sonication step allows the separation between intact crystalline domains, leading to isolated CNC bearing half-sulfate ester groups on their surface. These charges come from the reaction between sulfuric acid and surface hydroxyl groups of cellulose and induce repulsive forces between negatively charged CNC leading to colloidal stability and dispersion in water [15]. While sulfuric acid is the most common acid used for cellulose fibers hydrolysis, other researches have focused on the use of other organic or mineral acids, like phosphoric acid, hydrobromic acid, or hydrochloric acid [16–19], generally leading to less stable suspension due to the lack of charges at the surface of the CNC. Moreover, the use of deep eutectic solvents is the subject of the next part of this chapter. When

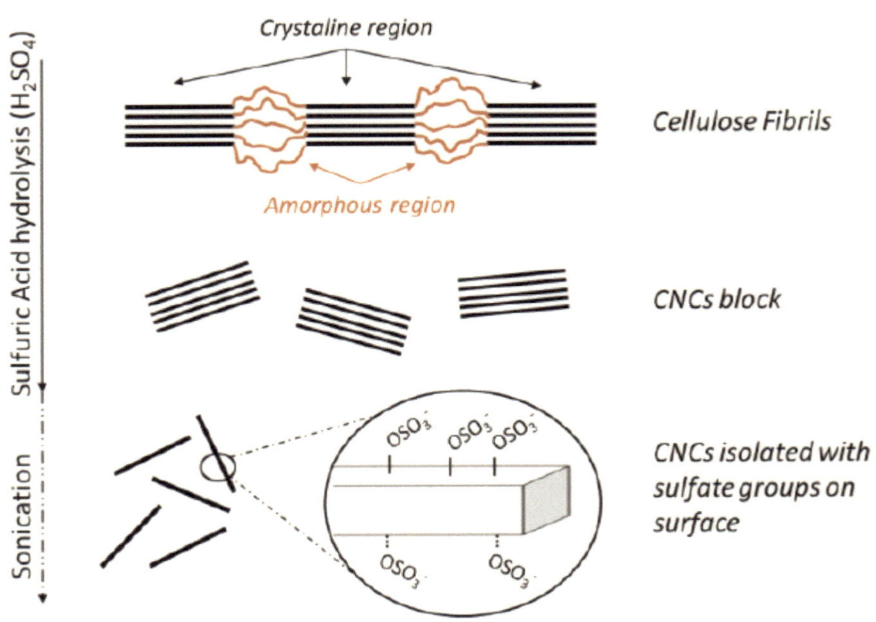

Figure 5.
Schematic representation of sulfuric hydrolysis of cellulose fibers (This scheme was extracted from an unpublished work (PhD manuscript of E. Gicquel, 2017)).

cellulosic source is not totally pure, previous steps are required. Indeed, alkali treatment (generally NaOH) and bleaching steps (generally acetic acid, aqueous chlorite) are essential to remove impurities, especially lignin and hemicelluloses when starting directly from biomass or even biowaste. Cellulose content of raw material is thus drastically increased. Note that a lot of studies have investigated the production of CNC from less conventional sources like rice, soy, and others in order to valorize food and organic waste [20, 21]. Final yield and morphology of CNC are really dependent on the cellulosic source and on the hydrolysis conditions. Indeed, optimization and control of the acid hydrolysis have been the subject of several publications. If common parameters are the hydrolysis with 64 wt% sulfuric acid at 40–45°C during about 30 min, it has been proved that variation of one of the parameters can largely influence the reaction yield as well as CNC properties. For example, by increasing of 10 min the time of hydrolysis, it has been shown by Flauzino Neto et al. that crystalline parts are destructed inducing a significant decrease in length [20]. Beck et al. [22] have confirmed this point, admitting that too long times of reaction induce degradation of cellulose but that too short times induce only large and non-dispersible fibers and large aggregates. Only specific reaction times yield to a well-dispersed colloidal suspension of CNC. Chen et al. [17] have confirmed that best yield and CNC properties are obtained with previously mentioned standard conditions. Moreover, the importance of acid concentration relative to cellulose fibers is highlighted too, since a too high concentration could be too drastic and a too low concentration insufficient for the hydrolysis efficiency. At the end of the reaction, mixture media are first diluted with distilled water to quench the reaction, then submitting to several separation steps with centrifugation cycles and filtrations and washed by dialysis against distilled water for several days, in order to remove unreacted compounds and chemicals. In some cases, they use also NaOH to neutralize pH which can modify the crystallinity and the surface ions. After dialysis, a final centrifugation cycle or another filtration process aims to remove aggregates. CNC suspension is finally sonicated in order to well disperse the nanocrystal and obtain colloidal suspension thanks to dimensions and sulfate half ester groups bearing by CNC.

2.2.2 Properties of cellulose nanocrystals

In addition to their nanometric size, CNC are unique biodegradable and renewable nanomaterials. Moreover, they result from a previously described optimized acid hydrolysis applied on abundant sources of cellulose and exhibit many other interesting properties. **Figure 6** summarizes the main CNC properties as well as their applications. Regarding the surface properties of CNC, they generally exhibit half-sulfate ester groups ($—SO_3^-$) on their surface after being treated with sulfuric acid. Even if the amount of $—SO_3^-$ is pretty low (about 50–200 μmol g^{-1}), these negative charges are sufficient to induce repulsive forces between nanomaterials and thus colloidal stability in aqueous media. Moreover, as presented in **Figure 6**, due to isolation process, other charged groups can be present on CNC surface, like carboxyl groups ($—COO^-$), aldehyde groups ($—CHO$), and others [23], leading to different charge properties inducing different CNC properties. Moreover, numerous hydroxyl groups (three groups in each AGU units) are at reactive surface sites for the introduction of new functional groups via hydroxyl groups' functionalization. Regarding the physical properties of CNC, they have a low density (1.606 g cm^{-3}), a high aspect ratio length/width (e.g., varying between 10 and 30 for CNC extracted from cotton and around 70 for those extracted from tunicate [15]), and a high surface area (between 150 and 800 m^2 g^{-1}). Note that all their morphological and surface properties are highly dependent on their source as well

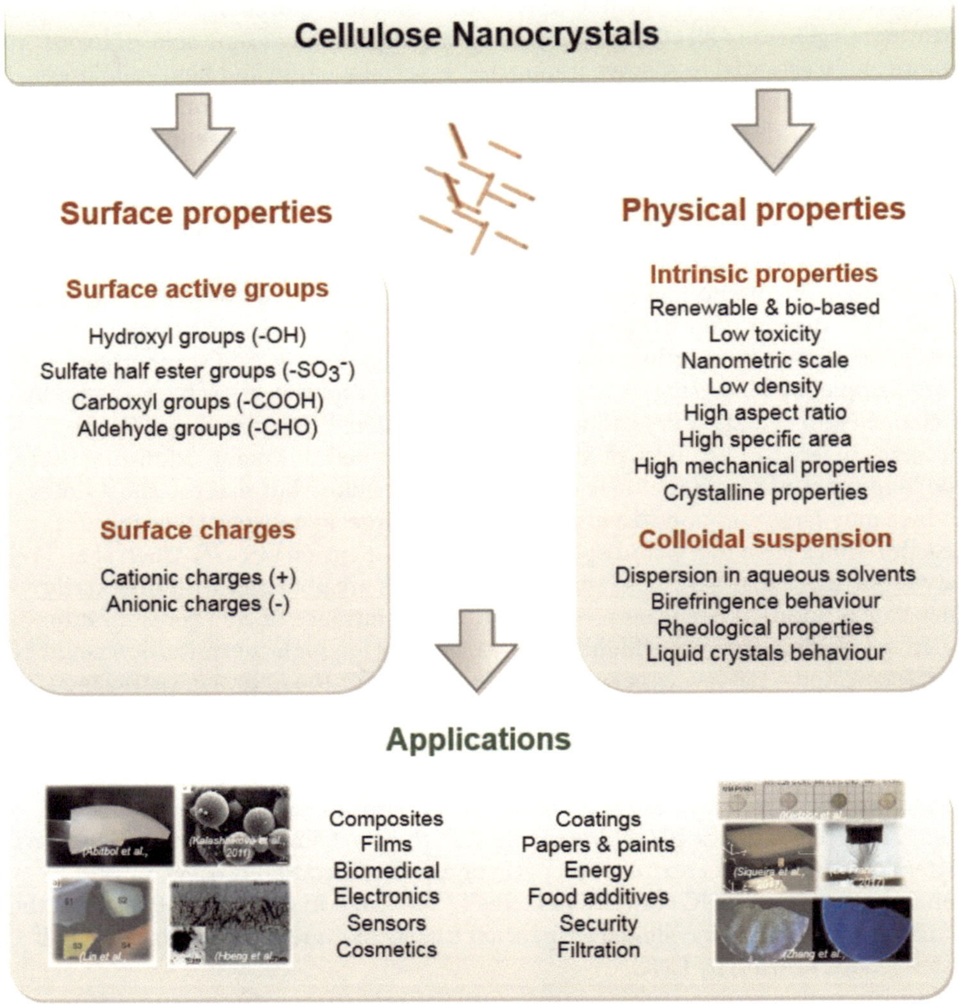

Figure 6.
Main surface and physical properties of cellulose nanocrystals and inherent main applications.

as their isolation process and conditions [8, 22]. Moreover, CNC exhibit highly interesting mechanical properties. Indeed, in addition to their high crystallinity (between 54 and 88% according to the source [24]), their high elastic modulus ($\approx$150 $\pm$ 50 GPa) and tensile strength ($\approx$7.5 $\pm$ 0.5 GPa) [25] make them interesting materials as mechanical reinforcement in polymer matrices, for example. For comparison, their mechanical properties are similar to Kevlar® fibers [26] widely used in composite field.

At low solid content (<3 wt%), due to hydrogen bonds between cellulose chains and thus between each nanocrystals, CNC water suspension is in the form of a translucent gel. Rheological properties of CNC are outstanding and concentration dependent. Indeed, at low concentration (<3 wt%), CNC suspension presents shear thinning behavior at high shear rate, and at higher concentration (>3 wt%), the suspension presents shear thinning behavior explained by the nanocrystals alignment in the flow direction at a critical shear rate [27]. Source and isolation of CNC influence these rheological properties too. Besides all these properties, CNC self-organize in ordered structure, especially to form a nematic phase. Revol et al. [28] described in the 1990s this self-organization of CNC in water suspension into stable chiral nematic phases. These last exhibit liquid crystalline properties, which when added to intrinsic birefringence of CNC induce interesting optical properties. Moreover, when ordered CNC suspension is solvent evaporated and thus solidified in a self-standing film, conserved chiral nematic structure (helicoidal structures) and iridescent behavior of films are observed and monitored by CNC concentration and surface charge as well as suspension sonication [27, 28]. **Figure 7** shows explicit pictures of these rheological and liquid crystalline properties.

All these outstanding surface and physical properties of CNC confirm their high and increasing interest in research and industrial field during the last decades. Although their isolation and characterization are currently well-advanced and optimized, application fields are at the center of ongoing researches, as described in the following part.

2.2.3 Various applications of cellulose nanocrystals and their industrialization

As exposed in **Figure 6**, CNC found applications in various fields. Indeed, thanks to their outstanding morphological, mechanical, and rheological properties as well as their colloidal stability and high surface reactivity. All these properties added to their biodegradability and renewability make them highly interesting and innovative materials with many potential applications. **Table 2** summarizes CNC applications and corresponding exploited properties, as well as some literature references.

Nanocomposite field is an emerging research area which finds applications in several domains like food packaging, medical devices, and building. Renewable aspect of CNC is particularly interesting since it correlates with the development of bio-based and biodegradable polymers as mentioned in the first part of this chapter. Moreover, these same properties are just as interesting in other application fields, from coatings, electronics, filtration, and biomedical devices to energy, cosmetics, and security. Note that for applications that may enter in contact with food or human body and for any industrialization, toxicity of CNC is a key challenge to investigate. Indeed, even if cellulose is known to be a nontoxic polymer, CNC are nanomaterials—and the "nano" prefix can be frightened for media and population

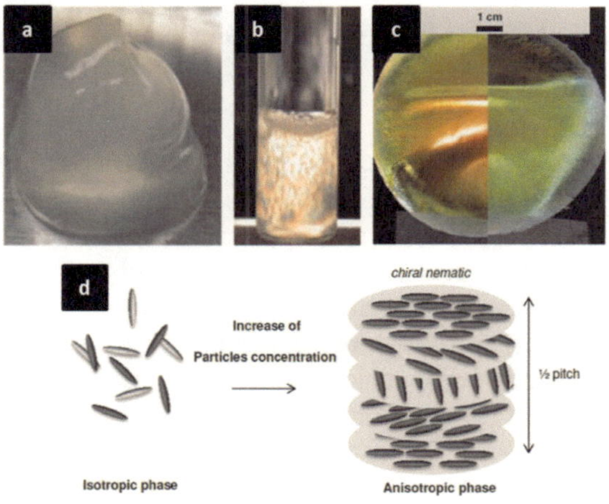

Figure 7.
(a) Translucent gel-like CNC suspension at 15 wt% in water (extracted from 57), (b) birefringence with shear-inducing observed for an aqueous CNC suspension at 0.6 wt% in cross-polarized light (extracted from [57]), (c) solvent-casted CNC film in diffuse light, normal to the surface (on the left part) and oblique to the surface (on the right part) (extracted from [29]), and (d) schematic representation of CNC orientation in isotropic and anisotropic phases (This scheme was extracted from an unpublished work (PhD manuscript of R. Bardet, 2014)).

Market	Applications	Exploited CNC properties	References
Composites/films	Nanocomposites Flexible packaging Optical films	High mechanical properties Filmogenic properties Morphology	[30, 31]
Coatings/paints/ adhesives	Coatings for flexible packaging	Morphology Rheological properties	[32, 33]
Electronics/sensors	Flexible electronics E-paper Piezoelectric sensors	Electrical insulating Piezoelectric properties Surface area	[34–36]
Filtration	Mesoporous films and membranes	High specific surface area High mechanical properties Hydrophilicity	[37]
Biomedical	Biocomposites for bone/tooth replacement Drug delivery Protein immobilization Wound dressings Biosensors	Low toxicity Colloidal stability High mechanical properties Surface reactivity	[38, 39]
Energy	Supercapacitors Flexible batteries	Strength Large surface area	[40]
Cosmetics	Hydrogels and foams	Colloidal stability Emulsion interfacial stabilization	[41, 42]
Security	Security papers and inks	Iridescent properties Morphology	[43]

Table 2.
Main market applications of CNC and corresponding properties and literature references.

—with specific morphology and surface properties. A recent review of Roman et al. [44] explored CNC toxicity. Results of this study correlate with previous results of Lin et al. [39] and Kovacs et al. [45], demonstrating that CNC are not toxic by ingestion or dermal contact and for aquatic organisms. However, pulmonary and cytotoxicity are less ideal since their toxicity depends on CNC properties and form (especially if they are in powder form, since they are more volatile). In any case, toxicity of CNC is low, especially when they are in wet-state or in composites, films, or coatings, for example, not constraining development of new CNC-based products.

3. Deep eutectic solvent for cellulose nanocrystals preparation

As described in the first part of this chapter, the traditional methods used to obtain cellulose nanocrystal consist of strong acid hydrolysis, enzymatic hydrolysis, or oxidation reactions. These treatments allow to hydrolyzing the amorphous regions in cellulose chains, and they are often followed by mechanical or ultrasonic treatment to homogenize the suspension. However, these methods use toxic chemicals and are difficult to industrialize.

Type	Component 1	Component 2
I	Quaternary ammonium salt	Metal chloride
II	Quaternary ammonium salt	Metal chloride hydrate
III	Quaternary ammonium salt	Hydrogen bond donor
IV	Metal chloride hydrate	Hydrogen bond donor

Table 3.
The four DES families.

Deep eutectic solvents (DESs) are a new class of green organic solvents; they are in the continuity of molten salt and ionic liquid solvent, but they are less toxic and easier to use. A deep eutectic solvent is composed of at least two compounds, a Lewis or Brönsted acid and a base. According to its constituents, a DES can be classed in one of the four existing categories listed in **Table 3**.

The association of these compounds with a specific ratio forms an eutectic mixture with a melting temperature far below than of its constituents.

In the case of type III, it is accepted that the self-association occurs via hydrogen bonding interactions between the hydrogen bond donor and the hydrogen bond

Cellulose source	Pretreatment (DES, etc.)	Molar ratio	Time	Temp. (°C)	Yield (%)	Dimension (nm)	References
Dissolving pulp	ChCl:OAD	1:1	2 h	100	68	l = 390 ± 25 d = 13.6 ± 1.1	[49]
		1:1	2 h	120	73	l = 353 ± 16 d = 13.8 ± 0.7	
	ChCl:p–t	1:1	2 h	60	70		
	ChCl:levulinic acid	1:2	2 h	100	88		
Dissolving pulp	ChCl:OAD	1:1	30 min	100		l = 50–350 d = 3–8	[50]
Cotton fiber	ChCl:OAD	1:1	1 h	100	79.8	l = 194.1 d = 9.6 ± 2.9	[51]
	ChCl:OAD	1:2	1 h	100	80.0	l = 152.7 d = 6.1 ± 1.2	
	ChCl:OAD	1:3	1 h	100	81.6	l = 122.4 d = 4.7 ± 2.2	
Cotton fiber	ChCl:OAD *Microwave assisted*	1:1 *800 W*	3 min	80	74.2	l = 100–350 d = 3–25	[52]
		1:1 *800 W*	3 min	90	62.4		
		1:1 *800 W*	3 min	100	57.8	l ≈ 150 d < 17	
Bleached eucalyptus kraft pulp	ChCl: OAD + *catalysis, $FeCl_3.6H_2O$ (mmol/gDES)*	1:4 *0*	7 h	80	86	l = 5152 ± 3328	[53]
		1:4 *0.15*	6 h	80	73	l = 270 ± 92	
		1:4 *0.3*	6 h	80	71	l = 258 ± 54	
		1:1 *0.15*	6 h	80	88	l = 5726 ± 3856	
Bleached birch Kraft pulp	AH:glycerol	1:2	10 min	70		d = 5.7 ± 1.3	[54]
Empty fruit bunch	ChCl:citric acid	1:2	6 h	85		l = 25–37	[55]
Dissolving pulp	GH:APA	1:2	24 h	Room T	80	d = 5.6 ± 1.4	[56]

ChCl, choline chloride; OAD, oxalic acid dihydrate; t, toluenesulfonic; AH, aminoguanidine hydrochloride; GH, guanidine hydrochloride; APA, anhydrous phosphoric acid.

Table 4.
Overview of the different DES pretreatments tested for cellulose nanocrystals obtention.

acceptor. The strong hydrogen bonds between the different compounds prevent the crystallization of each product and decrease the melting point of the mixture below room temperature [46]. Type III DESs are easy to obtain by simply mixing the compounds with the right ratio at a temperature higher than the melting point during 1 h. The mixture obtained is homogenous and transparent and has a low vapor pressure. However, the main drawback of deep eutectic solvents is their price. Nevertheless, some studies show that it is possible to recycle them between

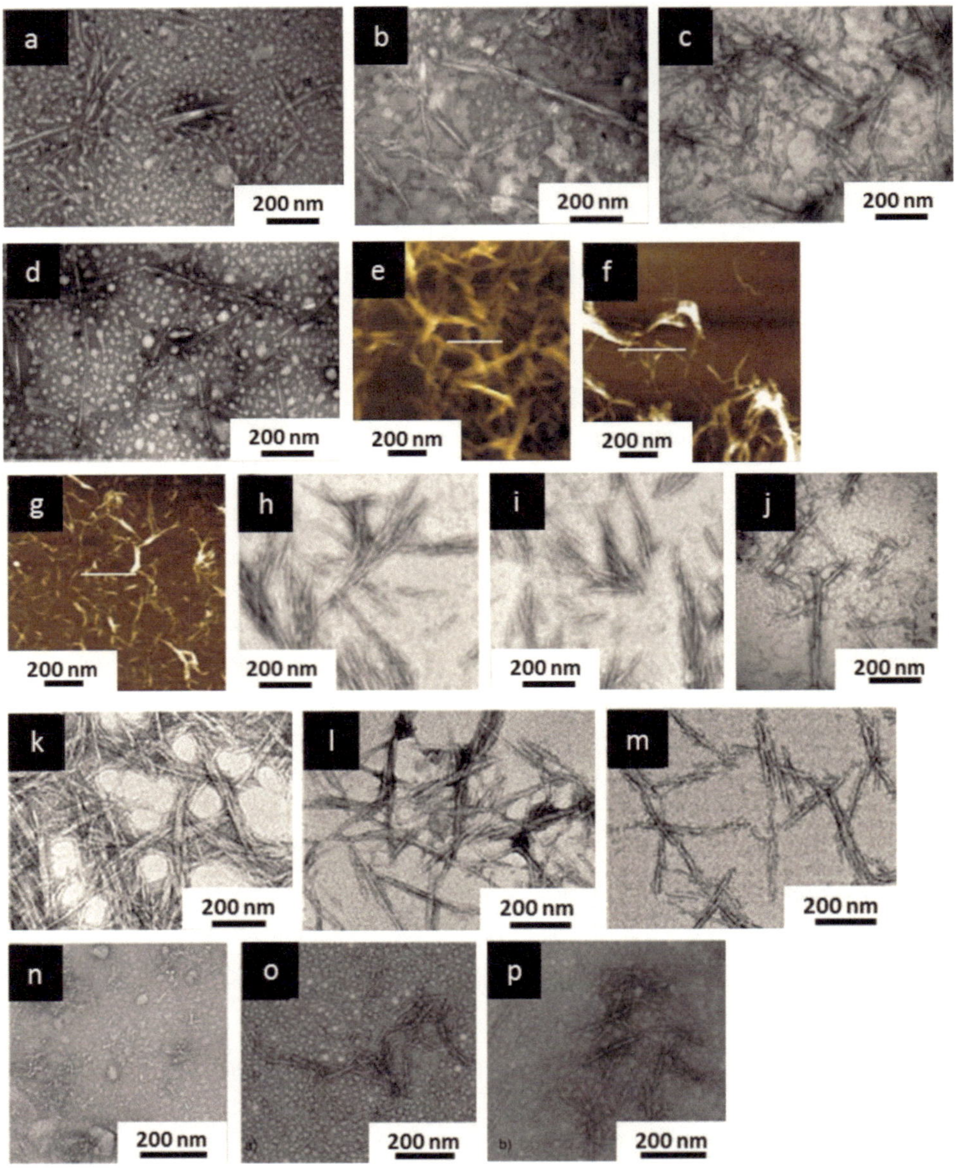

Figure 8.
Transmission electron micrographs of CNCs using DES pretreatment: (a) ChCl:OAD 1:1, 2 h–100°C [49]; (b) ChCl:OAD 1:1, 4 h–100°C [49]; (c) ChCl:OAD 1:1, 6 h–100°C [49]; (d) ChCl:OAD 1:1, 2 h–120°C [49]; (e) ChCl:OAD 1:1, 1 h–100°C [51]; (f) ChCl:OAD 1:2, 1 h–100°C [51]; (g) ChCl:OAD 1:3, 1 h–100°C [51]; (h) ChCl:OAD 1:4 + cat:$FeCl_3.6H_2O$ (0.15 mmol/gDES), 6 h–80°C [53]; (i) ChCl:OAD 1:4 + cat:$FeCl_3.6H_2O$ (0.3 mmol/gDES), 6 h–80°C [53]; (j) ChCl:OAD 1:1, 30 min–100°C [50]; (k) microwave-assisted ChCl:OAD 1:1, 3 min–80°C [52]; (l) microwave-assisted ChCl:OAD 1:1, 3 min–90°C [52]; (m) microwave-assisted ChCl:OAD 1:1, 3 min–100°C [52]; (n) guanidine hydrochloride:anhydrous phosphoric acid 1:2, 24 h room temperature [56]; (o) aminoguanidine hydrochloride:glycerol 1:2, 10 min–70° C [54]; (p) aminoguanidine hydrochloride:glycerol 1:2, 10 min –70°C [54].

three and five times depending on the DES components and the usage [47]. Moreover, a subclass of DESs, named natural deep eutectic solvents (NADES), is formulated using bio-based compounds; these solvents are environmentally friendly and their price can be lower.

DESs and NADES can be helpful in organic chemistry; indeed they can replace some toxic organic solvents; Zdanowicz summarized all their possible application domains for polysaccharide processing [48]. Among all these applications, one of them consists of the use of type III DESs as an acidic hydrolytic solvent for cellulose nanocrystal obtention.

This principle has been studied for the first time by Sirviö et al. [49]; they manage to extract individual nanocrystals using a choline chloride: oxalic acid dihydrate treatment from dissolving pulp. Different DESs had been prepared in different ratios and studied to hydrolyze the amorphous part of cellulose fibers from different sources: an overview of these treatments is summarized in **Table 4**.

In 2016, Sirviö tried three different DESs with choline chloride as hydrogen bond acceptor and oxalic acid (anhydrous or dihydrate), p-toluenesulfonic acid monohydrate, and levulinic acid as hydrogen bond donors [49]. Only DESs composed of choline chloride/oxalic acid dihydrate (ChCl:OAD) with a 1:1 molar ratio are allowed to obtain, after mechanical disintegration and CNC suspension. Different batches were prepared using different times and temperatures (**Figure 8(a–d)**). Cellulose nanocrystals obtained after 2 h of treatment at 120°C had the highest aspect ratio with a mean length of 353 ± 16 nm and diameter of 9.9 ± 0.7 nm (**Figure 8d**). The study showed that the final width of CNCs depends on the pretreatment temperature [57].

Ling et al. studied the effect of ChCl:OAD treatment on cellulose nanocrystal structure [51]; three molar ratios were chosen 1:1, 1:2, and 1:3, under two temperatures 80 and 100°C. Cellulose nanocrystals suspensions were obtained in every case (**Figure 8(e–g)**). Lower crystallinity and lamellar structures were observed for CNCs with lower acid content, and hydrogen bonds were more broken with higher acid ratio (ChCl:OAD = 1:3) during the DES treatment. Moreover, these CNCs obtained were better dispersed and had a higher aspect ratio.

In order to decrease the treatment temperature and increase the CNC yield, Yang et al. proposed to add a catalyst ($FeCl_3$) during the DES treatment [53]. They found out that the optimum conditions for the treatment were 80°C and 6 h using a deep eutectic solvent composed of choline chloride, oxalic acid dihydrate, and $FeCl_3{\cdot}6H_2O$ with a molar ratio of 1:4.43:0.1. Cellulose nanocrystals with a length between 50 and 300 nm and a diameter range of 5–20 nm were isolated from eucalyptus kraft pulp with a yield of 90% (**Figure 8(h–i)**).

Using ChCl:OAD (1:1) DES, Laitinen et al., in 2017, were able to obtain a CNC suspension after 30 min of pretreatment at 100°C and a microfluidizer treatment. The cellulose nanocrystals obtained had a low charged content and could be used as effective oil–water Pickering stabilizers (**Figure 8(j)**) [50].

In 2017, interesting work was published by Liu et al. [52]. They reported an ultrafast fabrication of CNCs using the DES ChCl:OAD (1:1) assisted by microwave pretreatment. In only 3 min, the CNCs obtained had diameters between 100 and 25 nm and lengths between 100 and 350 nm (**Figure 8(k–m)**). The yield was 74.2%, and the nanocrystals' crystallinity was higher than 82%.

Another DES composed of choline chloride and citric acid with a molar ratio of 2:1 allowed Ibrahim et al., in 2018, to hydrolyze lignocellulosic materials and to obtain cellulose nanocrystals [55].

Some deep eutectic solvents can dissolve cellulose; it is the case for the DESs composed of guanidine hydrochloride and anhydrous phosphoric acid (1:2). This solvent was studied by Sirviö to treat dissolving pulp during 24 h and then regenerate it into cellulose nanoparticles [56] (**Figure 8n**)).

In addition to the hydrolyzation of cellulose amorphous part, some DESs can simultaneously chemically modify the CNC surface. For example, Li et al. obtained cationic nanocrystals using aminoguanidine hydrochloride and glycerol mixture with a molar ratio of 1:2 (**Figure 8(o–p)**) [54].

4. Conclusion

The aim of this chapter was to provide an overview of cellulose nanocrystals and their production by various hydrolysis procedure.

Traditionally extracted from multiple cellulose sources via sulfuric acid hydrolysis, these cellulosic nanomaterials are pioneering for a wide range of applications, briefly introduced in this chapter. Such hydrolysis process has allowed the industrialization of this cellulose nanocrystals since 2011. However, in current context, the development of green chemistry is crucial in this research field, and new alternative process are expected.

In order to support this sustainable strategy, the use of natural deep eutectic solvents for the production of cellulose nanocrystals in mild condition has been very recently developed with interesting and encouraging results.

Acknowledgements

LGP2 is part of the LabEx Tec 21 (Investissements d'Avenir—grant agreement n °ANR-11-LABX-0030) and of PolyNat Carnot Institute (Investissements d'Avenir —grant agreement n° ANR-16-CARN-0025-01).

Conflict of interest

The authors declare no conflict of interest.

Author details

Manon Le Gars[1*], Loreleï Douard[1], Naceur Belgacem[1,2] and Julien Bras[1,2]

1 Université Grenoble Alpes, CNRS, Grenoble INP, LGP2, Grenoble, France

2 Institut Universitaire de France (IUF), Paris, France

*Address all correspondence to: manon.le-gars@lgp2.grenoble-inp.fr

References

[1] Varshney VK, Naithani S. Chemical functionalization of cellulose derived from nonconventional sources. In: Kalia S, Kaith BS, Kaur I, éditeurs. Cellulose Fibers: Bio- and Nano-Polymer Composites: Green Chemistry and Technology. Berlin, Heidelberg: Springer, Berlin, Heidelberg; 2011. p. 43–60. DOI: 10.1007/978-3-642-17370-7_2

[2] Suhas GVK, Carrott PJM, Singh R, Chaudhary M, Kushwaha S. Cellulose: A review as natural, modified and activated carbon adsorbent. Bioresource Technology. 2016;**216**: 1066-1076

[3] Klemm D, Heublein B, Fink H-P, Bohn A. Cellulose: Fascinating biopolymer and sustainable raw material. Angewandte Chemie International Edition. 2005;**44**(22): 3358-3393

[4] O'Sullivan AC. Cellulose: The structure slowly unravels. Cellulose. 1996;**4**:173-207

[5] Habibi Y. Key advances in the chemical modification of nanocelluloses. Chemical Society Reviews. 2014;**43**(5): 1519-1542

[6] Dufresne A. Nanocellulose: From Nature to High Performance Tailored Materials. Berlin: de Gruyter; 2012. 460 p

[7] Capron I, Rojas OJ, Bordes R. Behavior of nanocelluloses at interfaces. Current Opinion in Colloid & Interface Science. 2017;**29**:83-95

[8] Moon RJ, Martini A, Nairn J, Simonsen J, Youngblood J. Cellulose nanomaterials review: Structure, properties and nanocomposites. Chemical Society Reviews. 2011;**40**(7): 3941-3994

[9] Ranby BG. Fibrous macromolecular systems. Cellulose and muscle. The colloidal properties of cellulose micelles. Discussions of the Faraday Society. 1951;**11**:158-164

[10] Ranby BG, Ribi E. Über den feinbau der zellulose. Experientia. 1950;**6**:12-14

[11] CelluForce [Internet]. CelluForce. 2016. Available at: https://www.celluforce.com/en/technology/

[12] ISO/TC 6/TG 1-Cellulose Nanomaterials. Summary of International Activities on Cellulose Nanomaterials. 2017

[13] Cranston E. Overview of Canada. Chiba, Japan: TAPPI Nano Division; 2019

[14] Lima MM de S, Borsali R. Rodlike cellulose microcrystals: Structure, properties, and applications. Macromolecular Rapid Communications. 2004;**25**(7):771-787

[15] Habibi Y, Lucia LA, Rojas OJ. Cellulose nanocrystals: Chemistry, self-assembly, and applications. Chemical Reviews. 2010;**110**(6):3479-3500

[16] Camarero Espinosa S, Kuhnt T, Foster EJ, Weder C. Isolation of thermally stable cellulose nanocrystals by phosphoric acid hydrolysis. Biomacromolecules. 2013;**14**(4): 1223-1230

[17] Chen L, Wang Q, Hirth K, Baez C, Agarwal UP, Zhu JY. Tailoring the yield and characteristics of wood cellulose nanocrystals (CNC) using concentrated acid hydrolysis. Cellulose. 2015;**22**(3): 1753-1762

[18] Viet D, Beck-Candanedo S, Gray DG. Dispersion of cellulose nanocrystals in polar organic solvents. Cellulose. 2007;**14**:109-113

[19] Yu H, Qin Z, Liang B, Liu N, Zhou Z, Chen L. Facile extraction of thermally stable cellulose nanocrystals with a high yield of 93% through hydrochloric acid hydrolysis under hydrothermal conditions. Journal of Materials Chemistry A. 2013;**1**(12): 3938-3944

[20] Flauzino Neto WP, Silvério HA, Dantas NO, Pasquini D. Extraction and characterization of cellulose nanocrystals from agro-industrial residue—Soy hulls. Industrial Crops and Products. 2013;**42**:480-488

[21] Johar N, Ahmad I, Dufresne A. Extraction, preparation and characterization of cellulose fibres and nanocrystals from rice husk. Industrial Crops and Products. 2012;**37**(1):93-99

[22] Beck-Candanedo S, Roman M, Gray DG. Effect of reaction conditions on the properties and behavior of wood cellulose nanocrystal suspensions. Biomacromolecules. 2005;**6**(2): 1048-1054

[23] Tang J, Sisler J, Grishkewich N, Tam KC. Functionalization of cellulose nanocrystals for advanced applications. Journal of Colloid and Interface Science. 2017;**494**:397-409

[24] Lin N, Huang J, Dufresne A. Preparation, properties and applications of polysaccharide nanocrystals in advanced functional nanomaterials: A review. Nanoscale. 2012;**4**(11): 3274-3294

[25] Klemm D, Kramer F, Moritz S, Lindström T, Ankerfors M, Gray D, et al. Nanocelluloses: A new family of nature-based materials. Angewandte Chemie International Edition. 2011; **50**(24):5438-5466

[26] Cheng M, Chen W, Weerasooriya T. Mechanical properties of Kevlar® KM2 single fiber. Journal of Engineering Materials and Technology. 2005; **127**(2):197

[27] George J, Sabapathi S. Cellulose nanocrystals: Synthesis, functional properties, and applications. Nanotechnology, Science and Applications. 2015;**8**:45-54

[28] Revol J-F, Bradford H, Giasson J, Marchessault RH, Gray DG. Helicoidal self-ordering of cellulose microfibrils in aqueous suspension. International Journal of Biological Macromolecules. 1992;**14**(3):170-172

[29] Beck S, Bouchard J, Chauve G, Berry R. Controlled production of patterns in iridescent solid films of cellulose nanocrystals. Cellulose. 2013; **20**(3):1401-1411

[30] Mariano M, Kissi NE, Dufresne A. Cellulose nanocrystals and related nanocomposites: Review of some properties and challenges. Journal of Polymer Science Part B: Polymer Physics. 2014;**52**(12):791-806

[31] Hubbe MA, Ferrer A, Tyagi P, Yin Y, Salas C, Pal L, et al. Nanocellulose in thin films, coatings, and plies for packaging applications: A review. BioResources. 2017;**12**(1):2143-2233

[32] Kaboorani A, Riedl B, Blanchet P, Fellin M, Hosseinaei O, Wang S. Nanocrystalline cellulose (NCC): A renewable nano-material for polyvinyl acetate (PVA) adhesive. European Polymer Journal. 2012;**48**(11):1829-1837

[33] Mascheroni E, Rampazzo R, Ortenzi MA, Piva G, Bonetti S, Piergiovanni L. Comparison of cellulose nanocrystals obtained by sulfuric acid hydrolysis and ammonium persulfate, to be used as coating on flexible food-packaging materials. Cellulose. 2016; **23**(1):779-793

[34] Hoeng F, Denneulin A, Bras J. Use of nanocellulose in printed electronics: A review. Nanoscale. 2016;**8**(27): 13131-13154

[35] Csoka L, Hoeger IC, Rojas OJ, Peszlen I, Pawlak JJ, Peralta PN. Piezoelectric effect of cellulose nanocrystals thin films. ACS Macro Letters. 2012;**1**(7):867-870

[36] Gaspar D, Fernandes SN, De Oliveira AG, Fernandes JG, Grey P, Pontes RV, et al. Nanocrystalline cellulose applied simultaneously as the gate dielectric and the substrate in flexible field effect transistors. Nanotechnology. 2014;**25**(9):094008

[37] Karim Z, Mathew AP, Grahn M, Mouzon J, Oksman K. Nanoporous membranes with cellulose nanocrystals as functional entity in chitosan: Removal of dyes from water. Carbohydrate Polymers. 2014;**112**:668-676

[38] Domingues RMA, Gomes ME, Reis RL. The potential of cellulose nanocrystals in tissue engineering strategies. Biomacromolecules. 2014;**15**(7):2327-2346

[39] Lin N, Dufresne A. Nanocellulose in biomedicine: Current status and future prospect. European Polymer Journal. 2014;**59**:302-325

[40] Yang X, Shi K, Zhitomirsky I, Cranston ED. Cellulose nanocrystal aerogels as universal 3D lightweight substrates for supercapacitor materials. Advanced Materials. 2015;**27**(40):6104-6109

[41] Kalashnikova I, Bizot H, Bertoncini P, Cathala B, Capron I. Cellulosic nanorods of various aspect ratios for oil in water Pickering emulsions. Soft Matter. 2013;**9**(3):952-959

[42] Tasset S, Cathala B, Bizot H, Capron I. Versatile cellular foams derived from CNC-stabilized Pickering emulsions. RSC Advances. 2014;**4**(2):893-898

[43] Zhang YP. Nanocrystalline cellulose for covert optical encryption. Journal of Nanophotonics. 2012;**6**(1):063516

[44] Roman M. Toxicity of cellulose nanocrystals: A review. Industrial Biotechnology. 2015;**11**(1):25-33

[45] Kovacs T, Naish V, O'Connor B, Blaise C, Gagné F, Hall L, et al. An ecotoxicological characterization of nanocrystalline cellulose (NCC). Nanotoxicology. 2010;**4**(3):255-270

[46] Francisco M, van den Bruinhorst A, Kroon MC. Low-transition-temperature mixtures (LTTMs): A new generation of designer solvents. Angewandte Chemie International Edition. 2013;**52**(11):3074-3085

[47] Satlewal A, Agrawal R, Bhagia S, Sangoro J, Ragauskas AJ. Natural deep eutectic solvents for lignocellulosic biomass pretreatment: Recent developments, challenges and novel opportunities. Biotechnology Advances. 2018;**36**(8):2032-2050

[48] Zdanowicz M, Wilpiszewska K, Spychaj T. Deep eutectic solvents for polysaccharides processing. A review. Carbohydrate Polymers. 2018;**200**:361-380

[49] Sirviö JA, Visanko M, Liimatainen H. Acidic deep eutectic solvents as hydrolytic media for cellulose nanocrystal production. Biomacromolecules. 2016;**17**(9):3025-3032

[50] Laitinen O, Ojala J, Sirviö JA, Liimatainen H. Sustainable stabilization of oil in water emulsions by cellulose nanocrystals synthesized from deep eutectic solvents. Cellulose. 2017;**24**(4):1679-1689

[51] Ling Z, Edwards JV, Guo Z, Prevost NT, Nam S, Wu Q, et al. Structural variations of cotton cellulose nanocrystals from deep eutectic solvent treatment: Micro and nano scale. Cellulose. 2019;**26**(2):861-876

[52] Liu Y, Guo B, Xia Q, Meng J, Chen W, Liu S, et al. Efficient cleavage of strong hydrogen bonds in cotton by deep eutectic solvents and facile fabrication of cellulose nanocrystals in high yields. ACS Sustainable Chemical Engineering. 2017;**5**(9):7623-7631

[53] Yang X, Xie H, Du H, Zhang X, Zou Z, Zou Y, et al. Facile extraction of thermally stable and dispersible cellulose nanocrystals with high yield via a green and recyclable $FeCl_3$-catalyzed deep eutectic solvent system. ACS Sustainable Chemical Engineering. 2019;**7**(7):7200-7208

[54] Li P, Sirviö JA, Asante B, Liimatainen H. Recyclable deep eutectic solvent for the production of cationic nanocelluloses. Carbohydrate Polymers. 2018;**199**:219-227

[55] Ibrahim A, Abdullah MF, Sam ST. Hydrolysis empty fruit bunch (EFB) using green solvent. IOP Conference Series: Materials Science and Engineering. 2018;**429**:012059

[56] Sirviö JA. Fabrication of regenerated cellulose nanoparticles by mechanical disintegration of cellulose after dissolution and regeneration from a deep eutectic solvent. Journal of Materials Chemistry A. 2019;**7**(2):755-763

[57] Foster EJ et al. Current characterization methods for cellulose nanomaterials. Chemical Society Reviews. 2018;**47**:2609-2679

Chapter 6

Phytonanofabrication: Methodology and Factors Affecting Biosynthesis of Nanoparticles

Bipin D. Lade and Arti S. Shanware

Abstract

The greener way of producing silver nanoparticles is the easiest, cheapest and most efficient way of producing large-scale nanoparticles that have no adverse effect on the environment. The nanosynthesis using various methodologies and the biological synthesis of silver nanoparticles have been discussed in detail. Plant extracts have been known to be competent for the extracellular biosynthesis of silver nanoparticles suggested by the various publications. Further, effects of various sources and methods on nanoparticle synthesis have been examined. Additionally, the impact of conditions such as dark, light, heating, boiling, sonication, autoclave on the size and shape of colloidal nanoparticles has been analyzed. Moreover, effects of specific parameters such as leaf extract concentration, $AgNO_3$, reaction temperature, pH, light and stirring time for nanoparticle synthesis are discussed, and the impact of silver nanoparticles on plant physiology has examined.

Keywords: silver nanoparticles, plant extract, biological nanoparticle synthesis, reducing and capping agents

1. Introduction

Nanotechnology involves learning how to formulate a really Tiny substance, perhaps through a bottom-up or a top-down approach. The former bottom-up strategy uses the active participation of both the reducing and the capping agent to somehow normalize metallic salt in the creation of metallic nanoparticles. The actual entities that taken part could be electrons present in the functional group of the reducing/capping agent and the electrons of metallic ions, for example, silver salt. In such an instance, the electron is available in the reduction agent's outer shell that mostly reduces unstable silver or other metallic ions to form zero-valent stable silver or metallic nanoparticles. Another method is the top-down approach, where a certain bulk material structure is imposed under multiple external forces, strain, stress, physical pressure, tension, temperature, using devices, techniques, equipment and instruments that fracture bulk material into tiny nanosized nanorange particles [1].

2. Nanosynthesis

In nanosynthesis, the reactant A (metallic salts) interacts with the reactant B (capping agent) where the nucleation reaction occurs. The entire phenomenon occurs at a specific temperature, and the nanoparticle synthesis is significantly improved by the stabilizing (agent) to form metallic nanoparticles and by-products. In several cases, the by-product or larger reaction structures may lead to a mass agglomeration, which suggests aggregation by chemical bonding of larger particles. However, the reacting metallic salts and the capping agent form nanoparticles that have been stabilized by a capping agent that directly prevents mass synthesis.

2.1 Types of nanoparticle preparation

2.1.1 Physical preparation method

The physical method uses a "top-down" approach in which the bulk material is pulverized into fine particulate matter by means of force applied, such as crushing, impact, disruption, degradation, cutting, cryo-grinding, grinding, processing and homogenization. Physical methods use the milling process in micro-particle fracturing, and few examples involve ball milling, high-energy ball milling (HEBM), grinding, cryo-grinding, refining and homogenization: high process homogenization (HPH) and medium-pressure homogenization (UHPH). These physical productions are versatile methods that manufacture nanoparticles of larger size, diameter and volume, which are still controlled, generate surface defects, contamination, costly and time-consuming. It may also be possible to produce nanoparticles of the same size, but instrument assembly is very expensive and maintenance is too costly. This imperfection in the form of efficient and expensive maintenance culminated in the space travel of biological sources and methods for the nanoparticle manufacturing process [2]. The synthesis of nanoparticles using biological methods includes the synthesis of nanoparticles using living things or matter. There are several other ways to synthesize nanoparticles, such as chemical and biological methods.

2.1.2 Chemical preparation method

The chemical preparation process involves a bottom-up technique and is the most frequently employed method for preparing silver nanoparticles in water or organic solvents as a stable, colloidal dispersion. Various chemical preparation methods are available, such as sol system, chemical preparations: CdS base, CdS/HgS/CdS composite, sol-gel process, hydrothermal, pyrolysis and chemical vapor deposition (CVD) process [3] (metallic) silver salt to produce nanoparticles. Here, commonly used reductants are borohydride, citrate, ascorbate and elemental hydrogen that reduce (metallic) silver salt to produce nanoparticles. Chemically synthesized nanoparticles are made of Cu, Ag, Au, Gold, etc., although the use of toxic chemicals in the synthesis protocol cannot be avoided. Alternative methods for nanoparticle synthesis are therefore explored in order to minimize the use of chemicals, and several physical methods were developed and implemented for synthesis of nanoparticles to resolve the chemical preparation.

2.1.3 Biological preparation method

Physical methods are expensive and require strong material, toughness, strength, etc. for nanoparticle synthesis, which may not be suitable for large-scale

Sr. no.	Plant species	Plant part	NP size	References
1	*Aloe vera*	Leaves	50–350 nm	[5]
2	*Argyreia nervosa*	Seeds	20–50 nm	[6]
3	*Acorus calamus*	Rhizome	31.83	[7]
4	*Allium sativum*	Sucrose and fructose	4–22 nm, 4 4 ± 1.5 nm	[8]
5	*Boerhaavia diffusa*	Whole plant	25 nm	[9]
6	*Citrus sinensis*	Peel	10–35 nm	[10]
7	Cocos nucifera	Inflorescence	22 nm	[11]
8	*Calotropis procera*	Plant	19–45 nm	[12]
9	Olive extract	1 ml	30 nm	[13]
10	*Passiflora foetida*	Leaf disc	14 nm	[14]
11	*Terminalia chebula*	Fruit extract	100 nm	[15]
12	*Thevetia peruviana*	Latex	10–60 nm	[16]

Table 1.
Reports of silver nanoparticle synthesis using plant species, biomolecules, nanoparticle sizes and references [17].

production of nanoparticles. Although chemical methods are used in the mass production of nanoparticles, the key product generated in the synthesis of nanoparticles tends to cause pollution that ultimately poses a risk to life in terrestrial and aquatic life, leading to environmental accumulation. Physical and chemical methods therefore constitute invaluable processes, while the problem can be overcome by using a biological solution that uses living organisms, bacteria, plants, etc., which is sometimes called the green method. These green strategies include plants, microorganisms and enzymes that are considered to be nonhazardous to humans through the use of natural biological agents that are ecofriendly, biodegradable as reducing, capping and stabilizing agents for nanoparticle synthesis. These biological preparations require a living origin, such as insects, parts of plants, microorganism such as *F. oxysporum*, enzymes, amino acid, vitamin polysaccharides, polyphenols, amino acid, tea, coffee, wine, winery waste, banana, red grape, table sugar and glucose are used for capping and stabilizing. Plant components for the manufacture of nanoparticles have significant advantages over other methods such as easy preparation, processing, cost-effective, rapid development, reproducible, stable components, environmentally friendly, avoiding the use of harsh and toxic chemicals and zero contamination of the environment [4]. Nevertheless, plant-based preparation has become common due to its simple, comprehensive availability and extensive range of samples, including plant part extracts and natural products. In addition, plant, leaf core, root, latex, seed and stem contain polyphenols and have been successfully used for metal particle synthesis [4]. Moreover, plant extract, more specific leaf extract, seed extract, stem extract, fruit extract, root extract and flower extract are used for nanoparticle synthesis. **Table 1** displays many studies of plants mediated silver nanoparticles synthesis reports with their reducing biomolecules and the respective nanoparticle size.

3. Nanofabrication using plant sample

The biological method that uses plant samples is the easiest and cheapest way to use a bottom-up approach to nanoparticle synthesis. In broad aspect, there are large number of research papers representing other plant parts for nanoformulation, which includes callus, seed, fruit, stem, flower, tendril and root, which we termed as green synthesis because the initial as well as the byproduct after nanoparticle synthesis both are ecofriendly and biodegradable. This greener method of plant-derived reducing agent is used to reduce silver ions to form stable zero-valent silver called silver nanoparticles. Likewise, other metallic nanoparticles, such as gold, copper, silver, titanium, etc., are used for synthesis of nanoparticles using plant extract. The basic synthesis of nanoparticles using plant extract and silver salts was discussed in detail in the following paragraphs. Nanoparticle synthesis is a single pot reaction in which the preparation of the plant extract, $AgNO_3$, is performed and these solutions are combined to form silver nanoparticles [14]. Silver salt was usually used in this chapter for analysis along with plant extract for the formulation of nanoparticles.

3.1 Preparation of plant leaf extract

In general, the specific amount of the plant sample is determined in grams for the preparation of the extract in solvents, and the extract filtrate is diluted with sterile distilled water. The diluted filtrate is then used as a reducing agent for silver salts to form silver nanoparticles. The general procedure for the preparation of the extract is as follows: first, the 10 g leaf of Passiflora foetida washed with tap water and dried at room temperature. Second, washed dried leaf specimens chopped into small pieces and macerated directly into 75 ml of autoclaved distilled water. Eventually, the mixture is centrifuged at 4000 rpm for 10 min and filtered with the aid of muslin cloth or syringe filters (stored at 4 °C for future use). **Figure 1** shows the step-wise process of preparation of leaf extract by boiling method for the synthesis of silver nanoparticles [18]. As we are dealing with nanoformulation, it is very important to use the (diluted) plant extract at a minimum to reduce the amount of silver salt.

3.2 Preparation of $AgNO_3$ solution

Similar to plant extract, the amount of silver salt must be low and precise in order to form stable nanoparticles. If the reaction of silver salt is more or more in bulk, the synthesis of nanoparticles may be hindered. For this purpose, the minimum quantity of reacting material is used for the effective, slow, nucleation of nanoparticles. Because silver nitrate ($AgNO_3$) is very expensive, it is important to

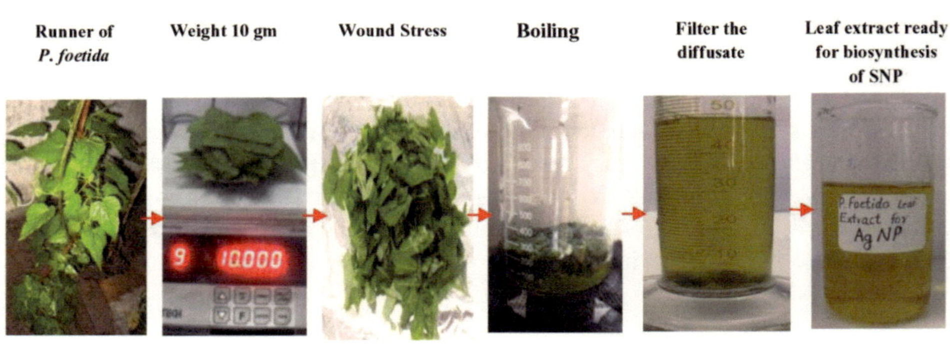

Figure 1.
Step-wise process of preparation of leaf extract of Passiflora foetida *by boiling method for the synthesis of silver nanoparticles [18].*

measure it in a very small quantity (1 mM) and therefore a stock of 100 mM aqueous silver nitrate solution ($AgNO_3$) is prepared and used to optimize the synthesis of silver nanoparticles.

3.3 Nanoparticle synthesis

Nanoparticles are easily synthesized by combining the appropriate concentration of plant extract and silver salt, and in short, 1 mM of silver nitrate (95 ml) is mixed with 5 ml of (diluted) leaf extract and kept at constant stirring in the 250 ml round bottom flask on a magnetic stirrer at room temperature. Later, solutions are allowed to react with each other in order to form nanoparticles. Synthesis of silver nanoparticles is confirmed by a change in the color of the solution from light green to dark brown, detected clearly by the naked eye (visual observation). This noticeable change in color is the primary indicator for the synthesis of Ag nanoparticles. The general optimization method for nanoparticle synthesis is shown in diagrammatic form in **Figure 2**.

Unless, after some time, the mixture does not show any change in color, it would be due to the acidity of the medium. It has been found in several papers that the alkaline solution is favorable to nanoparticle synthesis. Thus, 0.1 mM of NaOH is therefore used to produce a mixture solution into alkaline. It is vital that either the silver salt or the extract can be applied gradually using a syringe. The release of a moderate volume of extract into silver salt requires the proper synthesis of silver nanoparticles. The complete process of green synthesized silver nanoparticles at room temperature is shown in a diagrammatic form in **Figure 3**.

3.4 Nature of silver nanoparticles

Due to their application intent, colloidal nature of the synthesized silver nanoparticles is a critical aspect to consider. If the silver nanoparticles are not stabilized, they become heavy and will be found at bottom of flask or beaker. This aggregation problems can be resolved by sonicating the silver nanoparticle solution as well as using NaCl salt which hides the charges allowing the particles to clump together to form aggregates. Generally, a peak around 420–460 nm confirms stable silver nanoparticle synthesis, while silver nanoparticle aggregates form a broad peak around 350–525 nm with a decrease in the intensity of the plasmon absorbance [19].

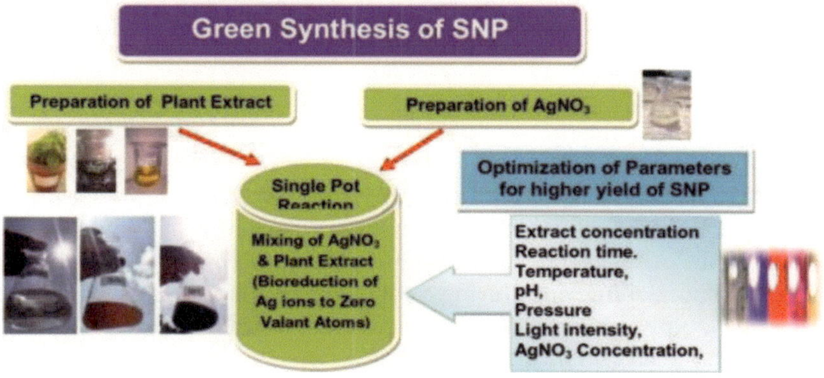

Figure 2.
General optimization method, criteria, instrument characterization and potential applications of synthesized nanoparticles [18].

Figure 3.
Complete process of green synthesis of silver nanoparticles at room temperature [18].

4. Nanoparticle synthesis methods and conditions

To nanotechnologists, nanoparticle formulation is being the most important aspect, where they use biological methods specifically plant extract to reduce metal ions and this will be used in further detail. The synthesis of nanoparticles is carried out using various methods such as microwave oven, autoclave, sonication, heating, boiling, etc., as well as under different conditions such as dark, daylight, light color different wavelength, etc. Likewise, the synthesis of nanoparticles by different methods implies the effect of specific system parameters such as light, heat, temperature, pressure, etc. on the kinetics of electrons present during nanoparticle synthesis in silver salts and capping agents. In comparison, the synthesis of nanoparticles in particular conditions indicates the effect of the conditions available on the synthesis of nanoparticles.

4.1 General synthesis process

In brief, the reactant, that is, plant extract and the $AgNO_3$ (1 mM) are mixed in a specific ratio such as 95:5 ml, respectively, and the reaction is conducted under specific instrument or machine, such as early mention of microwave oven, autoclave, sonication, heating, boiling, etc. During a study of these methods or conditions, the size or shape and type or form of nanoparticles formulated depend upon the kinetics of reaction that relays on the adopted methodology. In this section, we addressed different conditions under which nanoparticles could be generated. For example, the effect on the synthesis of nanoparticles of a specific sound wave may differ from that of light intensity or color. The sunlight has a specific effect on nanoparticle synthesis, which synthesizes nanoparticles of different sizes and shapes.

4.2 Various methods

4.2.1 Microwave oven

The microwave oven that could be used for synthesis of nanoparticles is one of the popular techniques. The microwave oven is a method for the processing of nanoparticles using different intensities of microwave radiation. Here the reactant, which may be a capping agent, is mixed with metallic salts and the mixture is placed in a microwave oven. Microwave treatment is administered to the reaction mixture in a short burst of 20 s for 5 min, which has influenced the kinetics of the mixture and drastically alters the size and shape of the nanoparticles. The radio wave in microwave oven is utilized to activate compounds or secondary metabolites in extract to reduce metal salts. Radio wave at a specific frequency is absorbed by extract components (phytochemicals and secondary metabolites) or activates certain extract enzymes to reduce metallic salts for stable nanoparticle production. As a result, microwave oven-assisted silver nanoparticles form in less time and are therefore the fastest way to synthesize [18].

4.2.2 Autoclave

The basic instrument used in the laboratory is an autoclave that produces steam at a pressure of 50 psi and reaches a temperature of ~121°C, at which organisms, pathogens and even endospores are killed or destroyed within 15 min. However, nanotechnologists use autoclaves to create a specific pressure and temperature for nanoparticle synthesis. In this process, kinetic of mixture is influenced by autoclave which is preheated for 5 min, once the pressure rises the flap is opened and the mixture flask or round bottom flask is kept in it for 5 min. Autoclave is not allowed to achieve 50 psi, instead, the autoclave is preheated to 5–10 psi for nanoparticle synthesis, and capping agent and the reducing agent are mixed with metallic salt and kept inside the autoclave for a specific time. Care must be taken in this method, as the autoclave is a very simple but dangerous machine, if the pressure rises above the crucial level and when it is opened critically results in steam or blast. Similarly, sonication is another way that could be used to synthesize nanoparticles [18].

4.2.3 Sonication

Sonication produces excitation of compounds from various biological sources, such as plants, microbes, animals, extracts, etc. It uses sound energy to agitate metallic salt in a reaction mixture ($AgNO_3$ + plant extract) to form nanoparticles. The specific wave hertz (Hz) is adjusted and applied to the reaction mixture for a specific time and the synthesized silver nanoparticles can be easily collected. Typically, an ultrasonic frequency of 20 kHz is administered to the reaction mixture. It is also referred to as ultrasound used for the synthesis of nanoparticles, nanoemulsions, nanocrystals and other forms. In addition, sonication is used for dispensing nanoparticles in liquid solution to disaggregate larger nanoparticles. Similar to sonication heating, the reaction mixture could also be another method for synthesis of nanoparticles [18].

4.2.4 Heating

In the heating process, heat energy is transferred to and from the mixture of extract and silver salt. Thus, when the reactants are heated, the heat excites the electron of the reacting secondary metabolites as well as the Ag^{2+} ions, the NO_3^{2-} ion silver salts to reduce Ag^{2+} ions to form Ag zero valent. In specific, the reaction mixture of metallic salt and plant extract is heated to a burner for particular temperature (10–100 °C) for certain period of time until the solution boils. Care must be taken not to overheat the reaction mixture as it comes out of the bottle. After boiling, the mixture of the solution may cool down at room temperature. To confirm that the synthesis of nanoparticles has been completed, the solution is observed to change the color from light green to dark brown. It may be possible to notice a change of color within minutes or it may take hours a day for a change of color depending on the reducing and capping agent involved in the process [18].

4.2.5 Boiling

The boiling method could be an approach in which the heat is given up to the boiling point in order to transform the liquid sample into a vapor state. Phytochemicals, secondary metabolites as well as metallic salt reactants are activated by boiling. The free electron of the secondary metabolite or phytochemicals reacts with metallic salt and helps to maintain stability. The reaction mixture of metallic salt $AgNO_3$ and plant extract is boiled on a burner or gas in a beaker for 5–10 min, the boiled solution that come out of a beaker, so care must be taken and the beaker is put out of a gas or a burner so that it may not falls out of a beaker. Once the solution boils, allow the mixture solution to cool down to reach room temperature and the color change from light green to dark brown is observed for verification of the synthesized nanoparticles [18].

4.3 Various conditions

The general condition that can be used in a laboratory is discussed in the following paragraph. In general, light of different wavelengths activates the enzyme and activated caused the reduction of silver or metal salts to the formation of nanoparticles and by-products. In this section, light-induced nanoparticle systems have been discussed.

4.3.1 Dark condition

Until going to light condition, it is very important to understand the effect of the dark condition on nanoparticle synthesis. Thus, the reaction mixture (extract + $AgNO_3$) is kept in a dark state. The nanoparticle synthesis reaction could be performed in dark where the conical flask is covered with aluminum foil to confer the darkness. Originally, the drops of $AgNO_3$ (1 mM) 5 ml are continuously applied to the known concentration of leaf extracts of a specific plant (*Passiflora. foetida*) 95 ml in conical flask. Such mixtures are kept at continuous stirring using a magnetic stirrer. The reaction must be conducted under dark conditions as light oxidized metal, resulting in photo-leaching due to its photo-reactivity properties. Some studies have indicated that synergistic antimicrobial activity is not linked to photo-activity. Nevertheless, the dark state of treated nanoparticles was found to be more stable than UV-light-synthesized particles [14, 18]. The [20] analysis is of synergistic bactericidal activity of silver titanium oxide ($AgTiO_2$) nanoparticles in both light and dark conditions. Therefore, the synthesis of nanoparticles using dark could be a convenient method in certain areas of application.

4.3.2 Sun light conditions

Corresponding to the dark environment, when the reaction mixture (extract + silver salt) is maintained under sunlight conditions, it may activate certain enzymes (phytochemicals and secondary metabolites) present in the leaf extract and express different results that are comparable to the reactions performed at a different condition/room temperature. There are several publications explaining the light-induced nanoparticle synthesis of silver, gold, titanium, aluminum, etc. The sunbeam emits electromagnetic energy in a wide range of wavelengths that enhances, activates or induces the kinetic reaction of silver salts with the functional group of the reacting secondary metabolites or phytochemicals present in the plant extract. Nanoparticles produced by sunlight are produced in much less time and are found to be more stable than room temperature and other conditions [14]. Apart from daylight, the other wavelength light could also be used for nanoparticle synthesis [18].

4.3.3 Light color (wavelength)

The light of different colors could produce different sizes and shapes of nanoparticles. Red, green, and blue are the primary colors of light that are used in different conditions in different ratios to create certain light colors. It is noted that silver nanoparticles under different light colors, such as green, blue, red, yellow and orange 15 W, result in silver nanoparticle synthesis where the blue color light of 15 W shows good UV absorption at 400 nm compared to other color lights. Therefore, the size and shape as well as the stability of nanoparticles could be altered by different wavelengths of light. Like the various wavelength of light, we would like to explore the idea of the influence of sound on nanoparticles [18]. **Figure 4** shows the silver nanoparticle synthesis using various methods and conformation by visual change in colors.

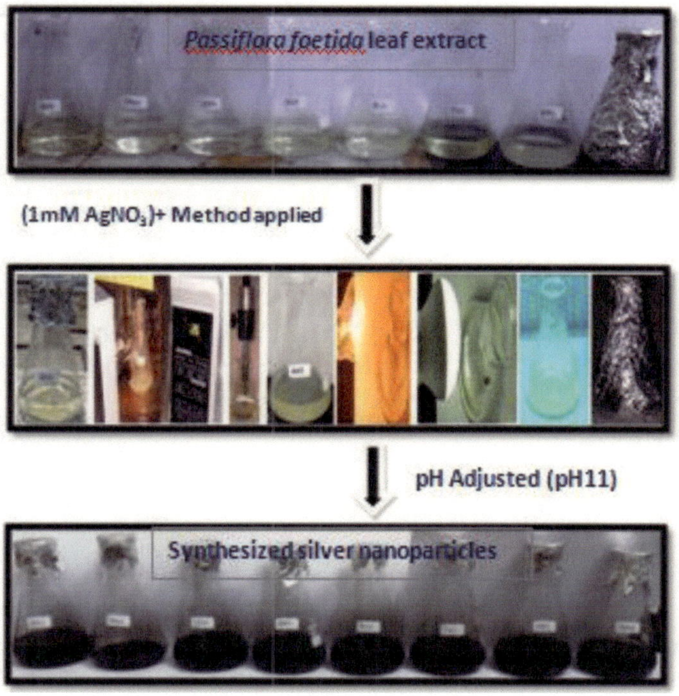

Figure 4.
Synthesis of silver nanoparticles using various methods/conditions: sun, microwave oven, sonication, autoclave, yellow bulb, white bulb, UV light and dark (from left to right) [18].

4.3.4 Influence of sound

In this process, the protocol is the same, and it is the reaction process carried under different conditions using selective instruments. The sound waves consist of longitudinal waves and transverse waves in solids that can be easily produced using a sound source such as the vibrating diaphragm of a stereo speaker. Therefore, the noise or acoustic environment is used to study different aspects of nanoparticles. Sound waves are used to isolate biological nanoparticles from bio-liquid samples. However, sound effects on the synthesis of bio-nanoparticles have not been studied. So, we are trying to provide some insight on the impact of sound on bio-nanoparticle synthesis. We are very eager to know the effect of sound including (jazz/rock/gaze/pop song) and the level of sound frequency on the synthesis of silver nanoparticles. Once the nanoparticles are synthesized, they are used for characterization by means of the available instruments, which will define nanoparticles size, shape, surface load, aggregates, colloidal, capping agent detection, reducing agent confirmation and conjugation (its ability to combine) with another polymer [18]. **Table 2** indicates the possible conditions under which the manufacture of silver nanoparticles is feasible.

4.3.5 Novel concepts

The novel concept could be traced out where different strategies can be applied to nanoparticle synthesis. It is reported in [14] paper that a specific type of wound stress has been applied to the plant leaf sample of the medicinal plant *Passiflora foetida* (leaf cut 2 cm in diameter and used to collect diffusate) and that secondary metabolite-induced stress has been extracted (see **Figure 5**). In this context, a stressed-induced plant extract is used, which contains stress-induced secondary metabolites and phytochemicals that reduce silver ion and act as a capping agent responsible for the synthesis of silver nanoparticles. The use of such stress-induced pool extracts may be considered for the formulation of silver nanoparticles [14].

Sr. no	Method/conditions	Point to note	Other remarks
1.	Dark condition	Light oxidized silver and leads to photo leaching	May cover with an aluminum foil
2.	Sunlight	Activate some enzyme present in leaf extract	Yields best results, Produced SNP quickly
3.	Domesticated microwave oven	Short burst of 20 s for 5 min to reaction mixture	Effects on SNP size and shape
4.	Autoclave	Preheating for 5 min once the pressure rises, the flap is opened and the mixture flask is kept in it for 5 min	If the pressure rises above the crucial level, opening it at that point may be critical
5.	Sonication	120 Hz of wave is adjusted and applied on reaction mixture for particular time	Varying Hz results in varied size and shape of SNP
6.	Light color	Green, blur, red, yellow and orange of 15 W	Watts could be changed for making various sizes/shapes of SNP
7.	Influence of sound	(Jazz/rock/gazel/pop song)	Volume level high/low may produce unique SNP

Note: The type or form of silver nanoparticles (SNPs) formulated depends upon the kinetics of reaction that relays on the adopted methodology.

Table 2.
Possible conditions under which the manufacture of silver nanoparticles is feasible [18].

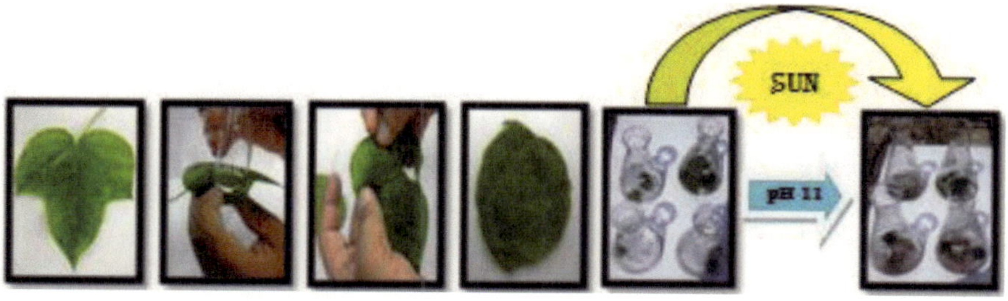

Figure 5.
Leaf disc preparation using test tube for applying wound stress to P. foetida *and synthesis of silver nanoparticles using leaf disc under sunlight [14].*

5. Factors affecting biosynthesis of nanoparticles

Manufacturing nanoparticles is a very critical process where the environmental factor is temperature and pressure; light intensity plays an important role in the synthesis of nanoparticles. Parameters such as sample leaf extract concentration, $AgNO_3$ concentration, reaction temperature, reaction time, pH reaction and different light reactions play a key role in the creation of silver nanoparticles of varying size, shape and dimension. Similarly, the light conditions such as sun light, blue light, red light, bulb light, tube light and in dark devoid of light effects are examined for nanoparticle biosynthesis. Several parameters that play a crucial role in the synthesis of the particular nanoparticles required have been addressed briefly. Such parameters are modified in order to produce the desired size of nanoparticles for the study of specific activity. Temperature is one of the most important parameters for nanoparticle synthesis.

5.1 Reaction temperature

Temperature has a specific effect on the process of synthesis of silver nanoparticles. Usually, the reaction is performed at room temperature, which takes a long time to complete, but can be accelerated by increasing the temperature of the reaction mixture. The temperature can be adjusted to between 30 and 100°C. Increased reaction temperature led to a rapid reduction in the rate of Ag^+ ions and subsequent homogeneous nucleation of silver nuclei, enabling the development of small size silver nanoparticles [18]. It has been found that as the temperature of the reaction mixture increases the rate of nanoparticle synthesis decreases, as well as the stability increases. In addition, silver nanoparticles synthesized at higher temperatures have small nanoparticle sizes. Another parameter that is important during silver nanoparticle synthesis is volume of leaf extract.

5.2 Volume of leaf extract

The quantity of leaf extract, that is, the volume of leaf extract, also affects the processing of nanoparticles and affects the time required for silver nanoparticle formation. Because leaf extracts are a major part of the reduction of silver ion, their volume up to a certain quantity is efficient in the formation of silver nanoparticles. Generally, the 1, 5, 10, 15, 20 and 100 ml leaf extract could be checked for effect. It is advised to use 1, 2, 3, 4 and 5 ml, since less volume is required for nanoscale particle synthesis. The average particle size of the synthesized silver nanoparticles is highly influenced by the concentration of leaf extract that the contents (phenol, polyphenols, polysaccharides, tannins and anthocyanins) contributes significantly to the reduction of Ag^{++} ions to the formation of stable Ag atoms in nanoscale, making it the main parameter during optimization for nanoparticle manufacturing.

5.3 Effect of pH

A major influence of the reaction (pH) is its ability to change the electrical loads of biomolecules, which may affect their ability to cap and stabilize and, consequently, the growth of nanoparticles [21]. In general, in the course of research work experimentation, our observations suggest that, if the pH of the reactant is acidic, the transition from light green to dark brown will take a much longer time, indicating that the acidic medium is not appropriate for nanoparticle synthesis. The size of nanoparticles is expected to be higher in the acidic medium than in the basic media, which is also reported in different studies. It was our observation that particle sizes at pH 3 were greater than those at pH 8 with regular spherical shape, both of which were observed in the TEM analysis. The alkaline pH condition facilitated the reduction and stabilizing capacity of antioxidants in the leaf extract. During optimization, the pH spectrum could be tested from 3 to 11 pH and potential spectroscopic UV tests could detect changes in absorption.

5.4 Silver nitrate ($AgNO_3$) concentration

The concentration of $AgNO_3$ is measured in a range of 0.1–1 mM or even more concentrations, such as 10, 50 and 100 mM [22]. However, 1 mM concentration is best studied and suggested as we are actually synthesizing nanoparticles that are very minute and cannot be seen with naked eyes. Therefore, a very small quantity of the reactant is required for the reaction to occur. If the concentration of the reactant is increased, the reduction of Ag^{++} will not be successful and the accumulation could be noticeable. Actually, it does not make sense to use higher concentration for synthesis of nanoscale particles. Another factor vital for synthesis is the reaction stirring time for up to which reaction mixture is stirred.

5.5 Reaction stirring time

The reaction stirring time is the time required for silver nanoparticle synthesis starting from the reactant is added in the beaker to occur reaction. The stirring time will enable the proper interaction of silver salt with the reducing complex components present in test leaf extract. The plant containing the more secondary metabolites or phytochemicals will reduce the silver salt in less time, in other terms the plant containing fewer reduced compounds will take longer time for reducing silver salt. However, the less number of secondary metabolites reduces silver salt and nanoparticles formed quickly in very less time. The stirring time will be dependent on reaction mixture acidity, basicity, temperature, reducing power of extract, light intensity, enzyme and secondary metabolites of the test plant extract. Ref. [23] concluded that the stability of silver nanoparticle synthesis using glutathione as a reducing agent increases at 72 h and confirms the visible UV absorbance at 344–354 nm. Thus stirring time affects the synthesis of nanoparticles, the duration of the synthesis of nanoparticles, the time allowed for the interaction of silver nitrate and leaf extract.

6. Fate of nanoparticles in plants

Targeted application nanoparticles have led to their use in many areas, including medical, pharmacological, chemical, paint, fertilizer, geosensing, agriculture, etc.

Nonetheless, nanoparticles are very durable and not easily degraded, which may remain in nature for decades or more. They may therefore remain in the climate, soil, air, water and eventually accumulate in the food chain. As a result, potential fates of nanoparticles in microorganisms, insects, animals, humans, plants and the atmosphere are very important.

6.1 Nanoparticles' effect on plants

Plants are faced with nanoparticles due to application in the field of plant protection. These nanoparticles are absorbed or collected or absorbed through the cell wall, the leaf surfaces, and the stomata or through the root from the soil. Once entry into the plant through the leaf surfaces, cell cytoplasm, mitochondria, ribosome, plant proteins, and enzymes radically alter their normal functions that cause cell death. They also interact with different processes in cells that cause alteration in phytohormones, metabolites, photosynthesis, transport and apoptosis-inducing metabolism. In addition, nanoparticles binding induce oxidative stress leading to degradation of proteins, lipids, nucleic acid, stress-related genes and increased antioxidant development for ROS activity, effects cell function and leads the oxidation of proteins, lipids and nucleic acid [24]. The nanoparticles could activate certain cells that activate apoptosis by intrinsic pathway or necrosis. Thus, while formulating the nanoparticles or the nanoparticles-based products for various applications, it is very important to keep the track of nanoparticle products side effect into environment. These nanoparticles interact through an unknown mechanism and sometimes act in support of and increase the growth of seeds and plants. Nanoparticles, on the other hand, could inhibit seed and plant growth without any clues. Nanobiopesticide (NBP) products must therefore be organic and nontoxic and their final deposition in plants must be studied exclusively. There are several studies that use nanoparticles or nanobiopesticides to combat pests; on the other hand, these nanobiopesticides are being purged into healthy plant cells. Therefore, the pesticide in nanoform or nanomaterial must be eco-friendly to plants and, at the same time, the NBP must act selectively to suppress pests or insects.

7. In vitro assay

The potential prospectus on the capacity of nanoparticles is tested via in vitro assay that based on Kirby-Bauer technique. The inherent potential of nanoparticles is used on various applications in laboratory as well as in industries. There are various ways to trace the inhibition property of silver nanoparticles. The minimum inhibition concentration of dried powder nanoparticles is important to determine the dose of nanoparticles against pathogens, and the zone of inhibition by silver nanoparticles is compared with the standard antibiotics. The concentration at which silver nanoparticles shows some effect to the microbe or fungi is determined by several methods such as agar well diffusion, disc diffusion and toxicity assays that are generally used for detecting the minimum concentration of silver nanoparticles for its antimicrobial/antifungal activity, larvicidal activity, anti-inflammatory, anti-platelet activity, anti-angiogenesis, antiviral, antilarva and pupicidal activity [25].

8. Applications

Nanotechnology has given nanomaterials a chance to be used in different fields. Thanks to their nanosize and stability, there are countless applications for nanoscale particles. Nanoparticles in nanoscale have a positive effect in their application in different fields, their modification at nanolevel in order to modify their properties and function to the advantage of food industries as well. The commercial ways have been taken up by the nanotechnology in food packing and processing. This technology has made possible to improve clinical clothing, plastic material, ceramic, surface disinfectant, dentistry purpose, virus, bacteria control, antimicrobial paints, textile, air disinfectant, waste water treatment, food preservation, agriculture application as pest, insect, larva, pupa control as nanoparticles pesticide (NPP), nanobiopesticides (NBP), nanofiler, nanofertilizer, etc. More than this, the plasma process is used in plasma farming, advanced packaging, labeling, sterilization and for food production applications. Therefore, the application of nanoparticles to artificial intelligence and robotics will contribute to the future of specific material benefit control for economic development.

9. Conclusion

Silver nanoparticle fabrication is a single pot reaction that transforms water soluble components such as $AgNO_3$, plant-based capping and stabilizing agent (secondary metabolites/phytochemicals) into water-insoluble components, that is, nanoparticles. Biological methods are the best methods due to their advantages such as low cost preparation, avoidance of environmental contamination, nontoxic by-products and adequate supply of samples, which can be scaled up for large production. Various conditions, such as dark, sunlight, microwave oven, autoclave, sonication, different wavelengths of light, heating, boiling, etc., and various other sources, methods, may apply as required by experiments. In addition, conditions such as dark, sun, heating, boiling, sonication and autoclave can be checked for the size and shape of colloidal nanoparticles. The stability of nanoparticles depends on different parameters such as leaf extract concentration, $AgNO_3$, reaction temperature, pH, light and stirring time to optimize and determine the size and shape of nanoparticles. Potential nanoparticles have been applied in many cross-disciplinary fields, from agriculture to medicine, but their effect, fate and accumulation on plants remain a mystery.

Acknowledgements

The authors would like to thank Rajiv Gandhi Biotechnology Centre, Rashtrasant Tukdoji Maharaj Nagpur University, L.I.T. Premises, Nagpur-440033 (M.S.), India, for providing research space and facility and Mrs. Diksha B. Lade, CHB Teacher, Department of Molecular Biology and Genetic Engineering, Rashtrasant Tukdoji Maharaj Nagpur University, Nagpur-440033 (M.S.), India, for manuscript editing.

Conflict of interest

The authors declare no conflict of interest.

Thanks

Author Bipin D. Lade would like to thank Dr. A. S. Patil and Dr. M. K. Rai and Department of Biotechnology, Sant Gadge Baba Amravati University, Amravati 444602 (M.S.), India, who introduced nanotechnology subject to Dr. Bipin D. Lade in initial career stage.

Author details

Bipin D. Lade* and Arti S. Shanware
Rajiv Gandhi Biotechnology Centre, Rashtrasant Tukadoji Maharaj Nagpur University, L.I.T. Premises, Nagpur, MH, India

*Address all correspondence to: dbipinlade@gmail.com

References

[1] Feynman R. There's Plenty of Room at the Bottom an Invitation to Enter a New Field of Physics. American Physical Society. Caltech's Engineering and Science; 1960 Available from: http://www.zyvex.com/nanotech/feynman.html

[2] Bharimalla AK. Chapter 4. In: Training-Manual Advances in applications of Nanotechnology, ICAR, Central Institute for Research on Cotton Technology (ICAR-CIRCOT) D.A.R.E, Ministry of Agriculture & Farmers Welfare, Govt of India. 2018. pp. 28-38

[3] Abegunde OO, Akinlabi ET, Oladijo OP, Akinlabi S, Ude AU. Overview of thin film deposition techniques. AIMS Materials Science. 2019;**6**(2):174-199. DOI: 10.3934/matersci.2019.2.174

[4] Kharissova OV, Dias HVR, Kharisov BI, Pérez BO, Pérez VMJ. The greener synthesis of nanoparticles. Trends in Biotechnology. 2013:240-248. DOI: 10.1016/j.tibtech.2013.01.003

[5] Chandran SP, Chaudhary M, Pasricha R, Ahmad A, Sastry M. Synthesis of gold nanotriangles and silver nanoparticles using aloe vera plant extract. Biotechnology Progress. 2006;**22**:577-583. DOI: 10.1021/bp0501423

[6] Thombre R, Parekh F, Patil N. Green synthesis of silver nanoparticles using seed extract of Argyreia nervosa. International Journal of Pharma and Bio Sciences. 2014;**5**(1):114-119

[7] Nakkala JR, Mata R, Gupta AK, Sadras SR. Green synthesis and characterization of silver nanoparticles using Boerhaaviadiffusa plant extract and their antibacterial activity. Industrial Crops and Products. 2014;**2**:562-566

[8] Ahamed M, Khan M, Siddiqui M, AlSalhi MS, Alrokayan SA. Green synthesis, characterization and evaluation of biocompatibility of silver nanoparticles. Physica E: Low-Dimensional Systems and Nanostructures. 2011;**43**:1266-1271

[9] Suna Q, Cai X, Li J, Zheng M, Chenb Z, Yu CP. Green synthesis of silver nanoparticles using tea leaf extract and evaluation of their stability and antibacterial activity. Colloids and Surfaces A: Physicochemical and Engineering Aspects. 2014;**444**:226-231

[10] Kaviya S, Santhanalakshmi J, Viswanathan B, Muthumary J, Srinivasan K. Biosynthesis of silver nanoparticles using *Citrus sinensis* peel extract and its antibacterial activity. Spectrochimica Acta. Part A, Molecular and Biomolecular Spectroscopy. 2011;**79**:594-598. DOI: 10.1016/j.saa.2011.03.040

[11] Mariselvam R, Ranjitsingh AJA, Usha Raja Nanthini A, Kalirajan K, Padmalatha C, Selvakumar MP. Green synthesis of silver nanoparticles from the extract of the inflorescence of *Cocos nucifera* (family: Arecaceae) for enhanced antibacterial activity. Spectrochimica Acta Part A: Molecular and Biomolecular Spectroscopy. 2014;**129**:537-541

[12] Gondwal M, Pant GJN. Biological evaluation and green synthesis of silver nanoparticles using aqueous extract of *Calotropis procera*. International Journal of Pharma and Bio Sciences. 2013;**4**(4):635-643

[13] Khalil MMH, Ismail EH, El-Baghdady KZ, Mohamed D. Green synthesis of silver nanoparticles using olive leaf extract and its antibacterial activity. Arabian Journal of Chemistry. Accepted Manuscript ARABJC 900. 2013:1-36. DOI: 10.1016/j.arabjc.2013.04.007

[14] Lade BD, Patil AS. Silver nano fabrication using leaf disc of *Passiflora*

foetida linn. Applied Nanoscience. 2017;**7**(5):181-119. DOI: 10.1007/s13204-017-0558-y

[15] Kiran Kumar HA, Mondal BK, Mohan Kumar K. Antimicrobial and antioxidant activities of Mimusopselengi seed extract mediated isotropic silver nanoparticles. Spectrochimica Acta Part A: Molecular and Biomolecular Spectroscopy. 2014;**130**:13-18. DOI: 10.1016/j.saa.2014.03.024.

[16] Rupiasih NN, Aher A, Gosavi S, Vidyasagar PB. Green synthesis of silver nanoparticles using latex extract of *Thevetia peruviana*: A novel approach towards poisonous plant utilization. Journal of Physics Conference Series. 2013;**423**:1-8

[17] Kavitha KS, Baker S, Rakshith D, Kavitha HU, Yashwantha Rao HC, Harini BP, et al. Plants as Green Source towards Synthesis of Nanoparticles. International Research Journal of Biological Sciences. 2013;**2**(6):66-76

[18] Lade BD. Biochemical and molecular approaches for characterization of wound stress induced antimicrobial secondary metabolites in *Passiflora foetida* linn [Ph. D thesis]. Amravati, MS, India: Biotechnology, Sant Gadge Baba Amravati University; 2017

[19] Lade BD, Gogle DP. Chapter 10: Nanobiopesticide: Synthesis and Applications in Plant Safety. In: Abd-Elsalam K, Prasad R, editors. Nanobiotechnology Applications in Plant Protection, The Nanotechnology in the Life Sciences. Vol. 2. USA: Springer Nature; 2019. pp. 169-189

[20] Li M, Noriega-Trevino ME, Nino-Martinez N, Marambio-Jones C, Wang J, Damoiseaux R, et al. Synergistic bactericidal activity of Ag-TiO_2 nanoparticles in both light and dark conditions. Environmental Science & Technology. 2011;**45**(20):8989-8995. DOI: 10.1021/es201675m

[21] Vadlapudi V, Kaladhar DSVGK. Review: Green synthesis of silver and gold nanoparticles. Middle-East Journal of Scientific Research. 2014;**19**(6):834-842. DOI: 10.5829/idosi.mejsr.2014.19.6.11585

[22] Gopinath K, Gowri S. Arumugam. Phytosynthesis of silver nanoparticles using *Pterocarpus santalinus* leaf extract and their antibacterial properties. Journal of Nanostructure in Chemistry. 2013;**3**:68. DOI: 10.1186/2193-8865-3-68.

[23] Balavandy SK, Shameli K, Biak DRBA, Abidin ZZ. Stirring time effect of silver nanoparticles prepared in glutathione mediated by green method. Chemistry Central Journal. 2014;**8**:11. DOI: 10.1186/1752-153X-8-11

[24] Rastogi A, Zivcak M, Sytar O, Kalaji HM, He X, Mbarki S, et al. Impact of metal and metal oxide nanoparticles on plant: A critical review. Frontiers in Chemistry. 2017;**5**:78. DOI: 10.3389/fchem.2017.00078

[25] Lade BD, Gogle DP, Lade DB, Moon GM, Nandeshwar SB, Kumbhare SD. Chapter 7: Nano-biopesticide formulations: Application strategies today and future. In: Nano-Biopesticides today and future Perspectives. Elsevier Academic Press; 2019. pp. 179-200

Chapter 7

The Components of Functional Nanosystems and Nanostructures

Gülay Baysal

Abstract

The science of nanosystems is used in many fields such as medicine, biomedical, biotechnology, agriculture, environmental pollution control, cosmetics, optics, health, food, energy, textiles, automotive, communication technologies, agriculture, and electronics. Nanomaterials, nanostructures, and nanosystems have recently brought the most popular and innovative approaches to our lives. This new technology is based on the production of invisible particles and the production of new materials by controlling the atomic sequence of these particles. Nanotechnological studies are based on mimicking the principle of atomic sequence in nature. Using a combination of different disciplines, it finds application in almost every field of our lives. Nanospheres, nanorobots, biosensors, quantum dots, and biochips are the main components of nanoparticles. Many new diagnostic and treatment methods are being developed nano-dimensional.

Keywords: nanospheres, nanorobots, biochips, nanosensors, biosensors, quantum dots

1. Introduction

Nanotechnology systems are rapidly becoming widespread as a technology for minimizing environmental damage and making materials advantageous. Nanotechnology has made significant improvements many application areas such as methods of obtaining solar power, the search for obtaining nuclear energy needs from solar energy, propulsion of water brought to fabrics, soil resistance, wrinkle resistance, bacterial escape, protection from UV rays, burning resistance, design of quantum computers, design of DNA computers speed, the creation of hydrogen fuel cells, the design of light space vehicles, the production of nano goods, biomolecules analysis, smart packaging in food packaging, nanosensors, and biosensors in the rapid diagnosis of diseases. **Figure 1** shows the structure of nanosystems.

2. The nanosystem unit components

2.1 Nanospheres and nanorods

Matrices, which have a particle size of 10–100 nm and which are prepared with natural or synthetic polymers and where the active substance is trapped in the particle, are called nanospheres. **Figure 2** shows the structure of the nanospheres. Polymeric nanoparticles are divided into nanocapsules and nanospheres.

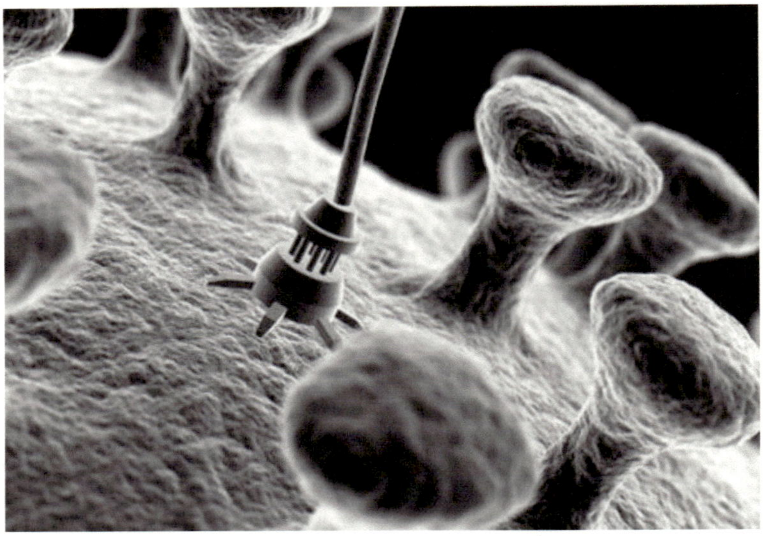

Figure 1.
Structure of nanosystems.

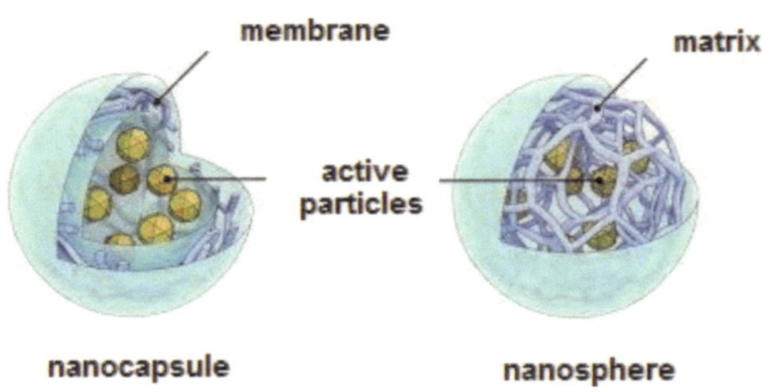

Figure 2.
The structure of nanospheres [1].

The nanospheres may be crystalline and amorphous. It is very advantageous because of its imprisonment, injection and dispersion properties [1].

Nanospheres have many advantages in their application areas. Nanospheres, especially in the field of medicine, can find a wide range of applications because they can easily pass through capillaries, penetrate tissue cavities easily, target organs, be formulated in a controlled manner, and reduce toxicity effects. Furthermore, due to the minimum particle size, they tend to agglomerate and form pellets and are challenging to obtain in liquid and solid forms. Nanospheres are obtained by polymerization, solvent displacement, salting out, controlled gelification, ionic gelation, solvent evaporation, and desolvation techniques [2–4] (**Figure 3**).

The nanorods are nanoscale materials that have unique optical properties like nanospheres. They are also expected to play an important role as both interconnects and functional units in fabricating electronic, optoelectronic, electrochemical and electromechanical nanodevices. Nanoscale materials such as fullerenes, nanorods, quantum dots and metallic nanoparticles have unique properties, because of their high surface area to volume ratio. Nanorods are used in many application fields such as dye solar cell, oligonucleotide detection, applied electric field and humidity sensitive. It should be recommended to synthesize nanorods in bulk in particular for biomedical fields. Nanorods are examined by extinction and scattering spectroscopy, DLS, TEM methods [5, 6]. Synthesis of gold nanorods has recently undergone dramatic improvements. It is possible to produce high yields of nearly

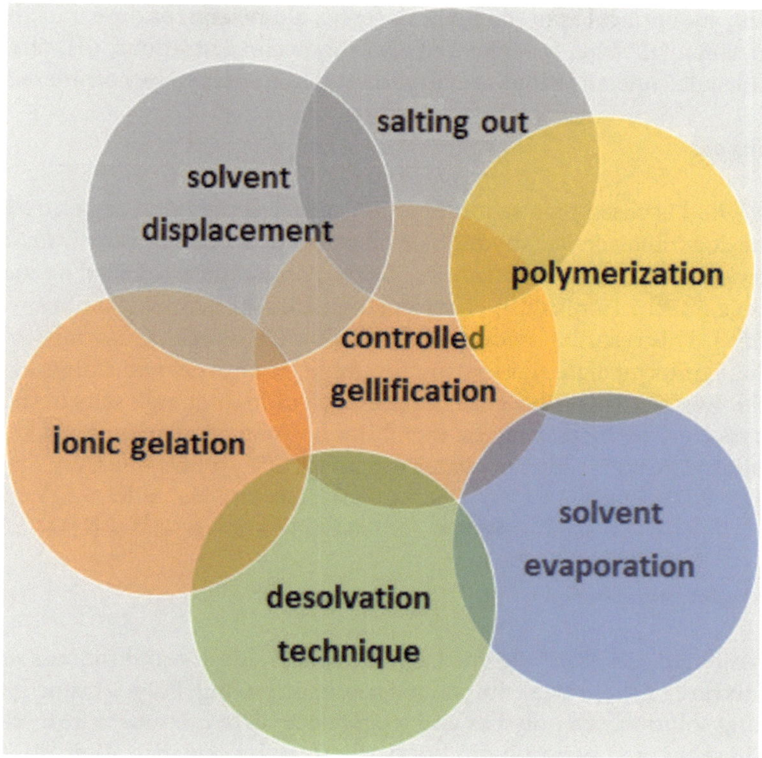

Figure 3.
The methods of obtaining nanospheres.

monodispersed short gold nanorods. The El-Sayed method of nanorod concentration determination is currently the standard way of measuring extinction coefficients (ε), and involves the coupling of bulk gold concentration, transmission electron microscopy (TEM) size analysis and absorbance data.

2.1.1 Polymerization

Emulsification polymerization is applied as a polymerization technique, and monomers are polymerized into nanospheres. Polymethylmethacrylate and poly cyanoacrylate polymers are among the most commonly used polymers for emulsification in the polymerization process. After polymerization, the active substance dissolves in the medium and is bonded with nanospheres or attached to the surface. The obtained material is purified, freeze-dried and the nanospheres are synthesized [7].

2.1.2 Ionic gelation

This method is based on the electrostatic interaction of the two aqueous phases and the nanospheres are prepared from natural polymers. Polyelectrolytes form hydrogel beads by cross-linking with counterions. These hydrogel beads are also called gelispheres. Gelispheres spreads into the polymer structure and forms a cross-linked cage. In this way, biomolecules can be trapped in this lattice structure. Examples of synthetic monomers and polymers used in this method include hydroxyethyl methacrylate, N-vinyl-2-pyrrolidone, N-isopropylacrylamide, vinyl acetate, acrylic acid, polyethylene glycol acrylate/methacrylate, polyethylene glycol diacrylate/dimethacrylate. Chitosan, gelatin, alginate fibrin, collagen, hyaluronic acid, dextran can be used as a natural polymer. Multivalent cations are generally magnesium, sodium, calcium, potassium, ferric, aluminum, barium and zinc ions. In the experimental stage, polymer and electrolyte concentrations, pH, temperature and biomolecule concentrations are important parameters to be considered.

2.1.3 Salting out

This method is based on a salt interaction such as a polymer, organic solvent, and magnesium chloride hexahydrate or magnesium acetate tetrahydrate. An emulsification mechanism is performed. Salting out method is based on increasing hydrophobic effect as a result of electronic repulsion of dissolved anions with high-density loads. This resulting hydrophobic effect increases the uniformity of the water-soluble intermediate phase structure, reduces entropy and causes agglomeration of the solvent. This is because the presence of high-charge salts in the system decreases entropy by increasing the regularity between similar surfaces, this result is also desirable. Anions reducing water solubility; $OH^- \approx SO_4^{2-} \approx CO_3^{2-} > ClO_3^- \approx Cl^- \approx OAc^- \approx IO_3^- > Br^- \approx I^- > NO_3^-$ and the cations: $Na^+ > K^+ > Li^+ \approx Ba^{2+} \approx Rb^+ \approx Ca^{2+} \approx Co^{2+} \approx Mg^{2+} \approx Fe^{2+} \approx Zn^{2+} \approx Cs^+ \approx Mn^{2+} \approx Al^{3+} > NH4^+ > H^+$.

2.1.4 Controlled gelification

In this method, gelation is formed using calcium chloride and sodium alginate. A suitable mixture of these two compounds results in gelling. Poly-L-lysine is added to the resulting solution as a polymer and a polyelectrolyte mixture is formed by mixing. Subsequently, nanospheres are synthesized by centrifugation [8].

2.1.5 Solvent evaporation

This method is based on the principle of emulsifying the active substance in the polymer and an organic solvent and removing the solvent by reducing the temperature and pressure. In this method, polyvinylchloride or gelatin may be used as the emulsifying agent.

2.1.6 Solvent displacement

It is one of the most widely used methods for the synthesis of nanoparticles. The solvent displacement is based on the displacement of a semi-polar solvent with the polymer interface. In this method, the organic phase containing the active substance and the polymer structure in the aqueous phase is self-emulsified. The polymer and active ingredient are dissolved in an organic solvent such as water-miscible ethanol, methanol or acetone. The organic phase is injected into the aqueous phase containing the active ingredient. The nanospheres are synthesized by precipitating the polymer in which the organic phase is dispersed in the aqueous phase.

2.1.7 Desolvation technique

This method is particularly preferred for obtaining nanospheres from natural polymers. The active substance is added to the polymer and solvent, and crosslinking is performed. Crosslinking agents must be added to effect crosslinking. The suspension is lyophilized by centrifugation, and the nanospheres are synthesized [9].

The most common and advantageous method used in the synthesis of nanospheres is the solvent displacement method. This method will provide great advantages especially in controlled drug release systems. Nanospheres synthesized by this method have more ability to convert poorly soluble and poorly absorbed drugs into better deliverables.

Particularly particle size and surface distribution properties are essential characteristic analyzes for the characterization of nanospheres. Because particle sizes and surface distributions also illuminate the properties of in vivo distribution, biological fate, toxicity and targeting ability. For the characterization of these properties, scanning electron microscopy (SEM), transmission electron microscopy (TEM), photon correlation spectroscopy methods are used. The Zeta potential analysis provides insightful information about load stability and particle collection. Zetasizer is used for this analysis. Fourier transform infrared spectroscopy (FTIR) analyses are performed to reveal the chemical bonding between the active substance and the polymers. The physical state of the active substance in the nanospheres is determined by differential scanning calorimetry (DSC) analysis after the lyophilization of the nanospheres. It can be applied many analysis methods such as in vitro drug release studies, drug release kinetics and stability study. Stability studies help to examine the effects of nanospheres on the physicochemical parameters of their formulations. In this way, suitable storage conditions are determined. Absorption and storage at room temperatures are performed for approximately 6 months and the observed changes in physicochemical parameters are recorded. Appropriate storage conditions are determined according to the results obtained. **Table 1** shows the intended use of the nanospheres synthesized according to the literature. **Figure 4** schematizes the synthesized gold nanospheres.

Nanospheres are used effectively in controlled drug delivery systems, tumor-targeted treatment methods, tumor-targeted treatment methods, epithelial cell therapy, genetic engineering studies, treatment methods targeting blood–brain barriers, cosmetics and many other areas [1].

2.2 Quantum dots

Quantum mechanics is the starting point of nanotechnology. These nano-sized semiconductor crystals are called quantum dots. Quantum dots are giant atomic structures that contain thousands of atoms. When substances are nano-sized, they

Purpose of usage	Nanospheres	References
Enhanced lithium storage properties	Ultra-small $ZnFe_2O_4$ nanosphere	[10]
Hydrogen production and photocatalytic activity	High specific surface area TiO_2 nanosphere	[11]
For lithium-ion batteries	Pyrite/carbon nanospheres	[12]
For sensing trace cysteine in HeLa cells	Gold-silver nanospheres	[13]
Water treatment and improved lubricating performance	Alkyl-capped copper oxide nanospheres	[14]
For PSA detection	Polydopamine nanospheres loaded with L-cysteine-coated cadmium sulfide quantum dots	[15]
For application in water treatment	MoO_3 nanospheres	[16]
For high-performance supercapacitors	CuS nanospheres	[17]

Table 1.
Some examples of nanospheres, according to the literature.

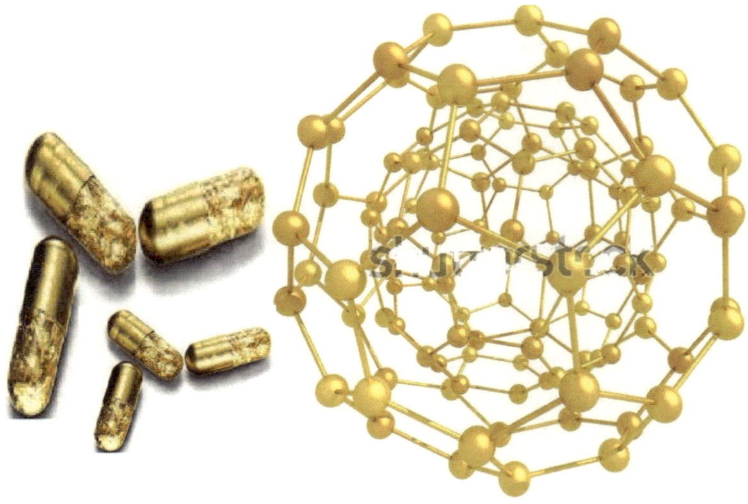

Figure 4.
The cancer treatment with gold nanospheres.

act according to quantum laws. The most preferred quantum points due to their semiconductivity, optical, and electrical properties are CdSe, InAs, CdS, GaN, InGeAS, CdTe, PbS, PbSe, ZnS. The controllable size of the quantum dots leads to outstanding optical and electrical properties, as the size of the quantum dots changes, the wavelength and color of their radiation changes. Quantum points are revealed by the stimulation of electrons [18, 19].

Quantum dots can be synthesized using methods such as plasma synthesis, viral coupling, bulk production, colloidal synthesis, fabrication, electrochemical coupling, and massive metal-free production. The parameters such as dimensions of quantum points, amount of solvent, amount of solution, amount of semiconductor metal, pH, and temperature are significant in the synthesis stage. **Figure 5** shows the wavelengths of quantum dots.

2.2.1 Fabrication

This method involves two different approaches, bottom-up and top-down. In the top-down method, small particles are formed by lithography. **Figure 6** shows quantum dots fabrication process.

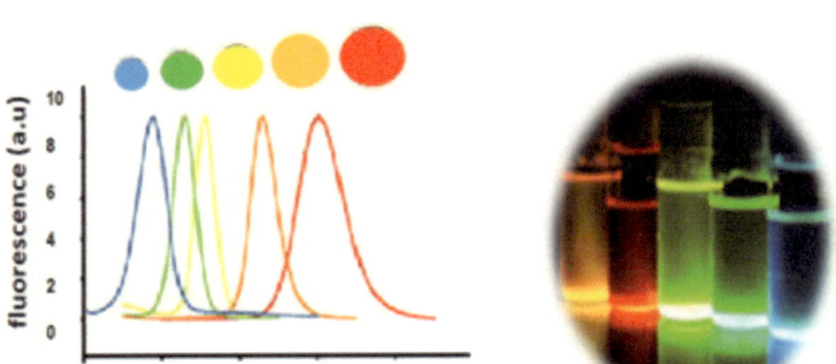

Figure 5.
The wavelengths of quantum dots [18].

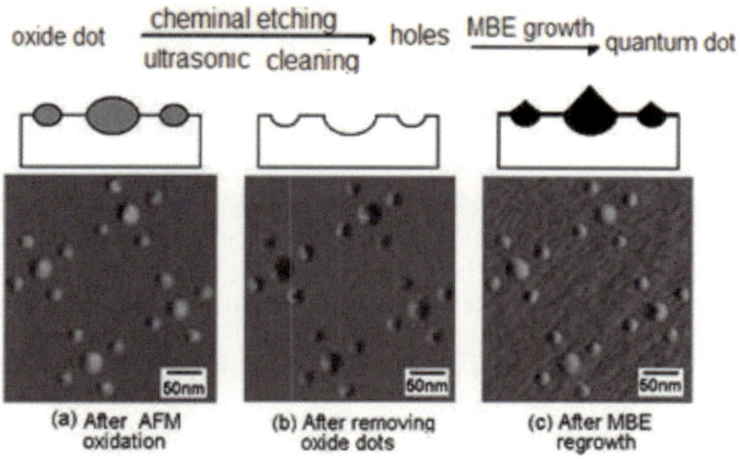

Figure 6.
Quantum dots fabrication process. (a) After AFM oxidation. (b) After removing oxide dots. (c) After MBE regrowth.

2.2.2 Colloidal synthesis

Colloidal synthesis is a practical synthesis technique where quantum dots can be synthesized easily under laboratory conditions. They consist of three main components: precursors, organic surfactants, and solvents.

2.2.3 Electrochemical coupling

Electrochemical coupling is a technique in which quantum dots can form spontaneously regularly. As a result of the ionic reaction at the electrolyte-metal interface, the nanostructures spontaneously form on the metal.

In the field of medicine, positron emission tomography and single-photon emission computed tomography are used in nuclear imaging systems, especially in the diagnosis of cancer diseases. Quantum dots can also be used in many engineering branches such as more efficient solar panels, bio-agents used for diagnostic purposes in medicine, low-energy lasers, LED lights of the desired color, low-energy, and more-lit bulbs, low-energy plasma televisions and displays. **Figure 7** shows the visualization of the quantum dots under UV light to detect different tumor cells by the addition of bioagents. Biological applications of quantum dots are examples of DNA protein sensors, sugar sensors, immunoassays, live cell imaging, bio-sensing, in vitro imaging, biological imaging, single molecule tracking, in vivo and animal

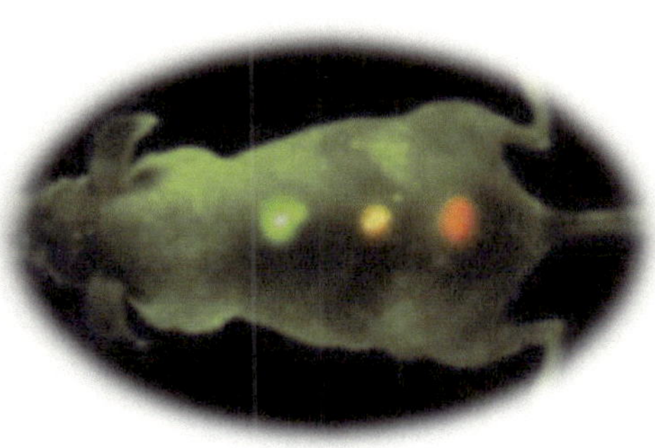

Figure 7.
Visualization of the quantum dots under UV light to detect different tumor cells by the addition of bioagents [19].

imaging. UV-vis and photoluminescence spectroscopy are generally used for characterization of quantum dots. Thanks to these methods, it is possible to perform fast, undamaged and contactless characterization. The use of photomodulated reflectance spectroscopy, an experimental method, offers a wide range of critical point advantages. The optical properties of quantum dots can be controlled by the dimensions of the dots. The size of the quantum dots can be characterized by scanning electron microscopy (SEM), transmission electron microscopy (TEM), and dynamic light scattering (DLS). Optical activity properties and particle sizes of quantum dots can be measured by photoluminescence excitation and Raman scattering spectroscopy. Atomic force microscopy (AFM), scanning tunneling microscopy (STM) and transmission electron microscopy (TEM) analysis methods are used to display the particle sizes. However, the best analysis results are obtained with AFM and TEM analysis methods.

3. Nanodevices and the complex nanostructures used in nanodevices

3.1 Nanorobotics

In order to develop nanorobots, firstly, the nanoparticles must be accurately and adequately analyzed, and the atomic arrays must be designed correctly. The essential components of nanorobots are nanocomputer. Especially in the field of health, it is vital for the rapid and accurate diagnosis and monitoring of diseases. In recent years, the need for diagnosis and treatment of cancer diseases, which is rapidly becoming widespread day by day, has significantly increased. Besides, there is an increasing need for nanorobots in environmental engineering, food technologies, gene therapy, machine design, military, and space science (**Figure 8**).

3.1.1 Biochips

As a product of the physical and biological sciences, biochips provide molecular analysis with high efficiency. Especially in gene engineering, biology, and medicine, biochips that do wonders in early diagnosis of diseases also enable gene sequencing, pharmacology and toxicology, analysis of DNA/RNA strands, and accurate identification of proteins. The most current studies are the determination of gene expression between human cells and tissue. In this way, global gene expression analysis is highly illuminating in the early detection of tumors in the living body. Application areas of biochips can be listed as a diagnostic tool in clinical medicine, quantifying

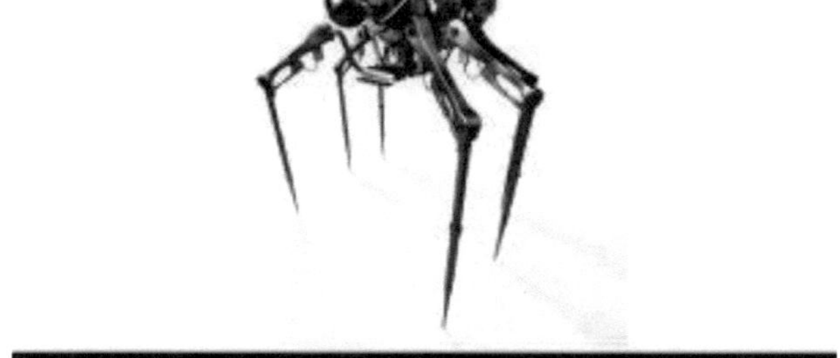

Figure 8.
The nanorobotics.

biomolecules, develop polymorphism analysis, cancer, rapid diagnostic testing, identifying, biowarfare agents [20].

Biochips provide great convenience in early and accurate diagnosis of diseases by reducing the detection time of protein sequences to less than 15 min and reducing the analysis time of nucleic acid sequences to less than 2 h. Especially in the food industry, DNA amplification of the genes of the target pathogen allows for rapid and accurate identification of pathogens. Biopharmaceuticals are also needed to identify mutations related to rifampin resistance in mycobacteria (RIF). Biochips are intended to be used to create a giant database for the narration of living and occurring events in the world soon. Biochips that can predict the health history of a person who is injured, sick or exposed to an accident, and that can give information about the food microbiota and nutritional values at every stage of food safety technologies from food production to consumption, and that can make diagnosis of diseases such as blood pressure and high blood sugar will be synthesized in the nearest future.

3.2 Nanosensors

Nanosensors are a combination of chemical, biological, and surgical sciences used to deliver nanoparticles to the world macroscopically. Nanosystems such as porous silicon, nanoparticles, nanoprobes, nanowires, nanotubes are widely used in the design of nanosensors. Examples of nanoparticles used in the design of nanosystems are MNPs magnetic nanoparticles, AuNPs gold nanoparticles, upconversion nanoparticles, QDs quantum dots, SWNTs single-wall carbon nanotubes, MWNTs multiwall carbon nanotubes, nano barcode technology and electronic nose [21]. These devices are tiny devices that can detect and respond to physical stimuli such as biological and chemical substances, displacement, motion, force, mass, acoustic, thermal and electromagnetic. In the literature studies, many nanosensors were synthesized for different purposes (**Table 2**).

Purpose of usage	Nanosensors	References
Detection of ochratoxin for real-time display of arsenic (As^{3+}) dynamics in living cells	Black phosphorene nanosensor (FRET)-based nanosensor	[22, 23]
Near-infrared imaging of serotonin release from cells	Fluorescent nanosensors	[24]
For improved dialysis treatment	microfuidic DNA-based potassium nanosensors	[25]
Development of ethanol and acetone gas sensing performance	$MgCO_2O_4$ nanosensor	[26]
Anticancer drug	SPR nanosensor	[27]
Detection of mercury ions	Multimodal nanosensor	[28]
Detection of serum albumin	Copolymer nanosensor	[29]
Detection of cysteine	Colorimetric nanosensor	[30]
Detection of cadmium ions	Quantum dots based-fluorescence nanosensörler	[31]
Emerging strategies	AuNP-based ICTS nanosensor	[32]
Detection of curcumin	Carbon-based chem nanosensor	[33]

Table 2.
Some examples of nanosensors, according to the literature.

Figure 9 shows the different uses of areas of nanosensors. The features to be considered in the design of nanosensors are selectivity, calibration requirement, reproducibility, stability, high sensitivity, wide measuring range, service life, the limit of determination, and sterilization (**Figure 10**).

Nanosensors consist of transducers and nanoparticles. Nanosensors can be designed as amperometric, voltammetric, potentiometric, colorimetric, SPR, fluorescence, optical fiber, SERS, acoustic and pieozoelectric transducer. Metallic, magnetic, quantum dots, graphene oxide, carbon nanotubes and upconversion nanoparticles are commonly used in nanosensors.

Xiang et al. [22] designed black phosphorene nanosensors for voltammetric analysis of ochratoxins. As known, ochratoxins have been identified as immunotoxic, nephrotoxic and carcinogenic in humans by the International Agency for Research on Cancer (IARC). Therefore, rapid and precise measurements are needed to determine whether ochratoxins are above the limit values in food components. Black phosphenes (BP) are widely used in the design of nanosensors because of their precise measurement capabilities. However, black phosphenes are highly reactive to water and oxygen, so they are easily affected by ambient conditions, so they lose their stable structure. In order to prevent this problem, the stability of BPs with covalent aryl diazonium, ligand surface coordination and coating materials is increased. The interaction of Ag^{+} ions and BPs in the N-methyl pyrrolidone (NMP) environment increases the super electrochemical properties considerably.

Soleja et al. [23] devised fluorescence resonance energy transfer (FRET) based nanosensors to detect arsenic metal and to determine its concentration. As it is known, arsenic is a toxic and heavy metal that has a carcinogenic effect and can cause serious health problems. Therefore, it is necessary to determine the concentration values accumulated in the living organism easily. The transcriptional repressor Arsr of the ars has an affinity for As^{3+} and the arsenic ions are thus more

Figure 9.
Different uses areas of nanosensors.

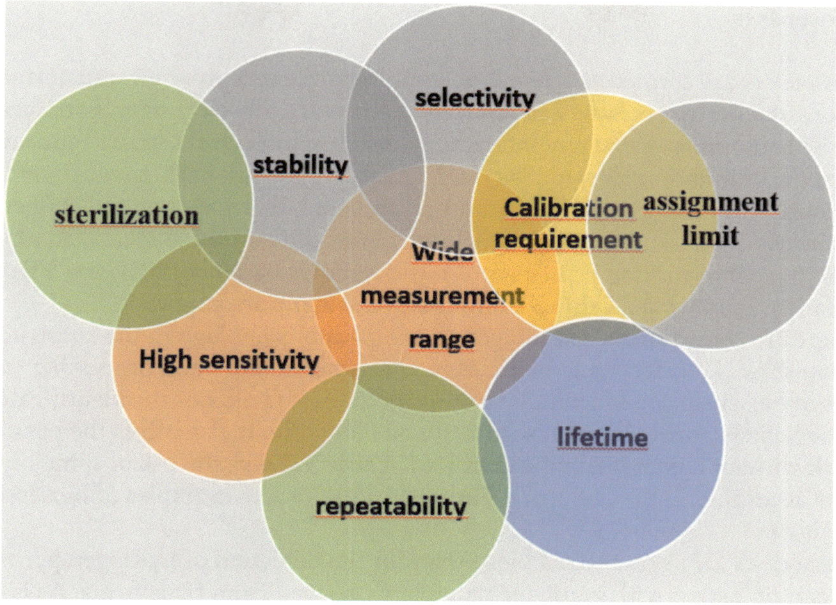

Figure 10.
The important properties of nanosensors.

easily bound. This process makes it possible to develop fluorescence resonance energy transfer (FRET) based nanosensors. The donor and acceptor perform ArsR binding with arsenic at the N- and C-terminus to obtain a recombinant protein.

Fluorescence sensors technology enables sensitive and non-destructive detection of signals in food additives and metal ions to detect environmental contamination. Fluorescence sensors have two important units: receptor and signal. As the target analyte concentrations are usually low, specific recognition for receptors is of great importance. Enzymes, aptamers and natural receptors, as well as artificial polymeric receptors or molecularly imprinted polymers, are used as receptors. Molecularly imprinted polymers are cross-linked polymers containing voids specific to analytes. These gaps support high selectivity. Comparing molecularly imprinted polymers and natural receptors, it is concluded that molecularly imprinted polymers show high chemical and physical stability, low cost and easy preparation properties. In fluorescence signal units, quantum dots, metal nanoclusters and organic dyes are often preferred as fluorescence sensors components.

Luo et al. [34] detected *E. coli* bacteria in milk by using radial flow chromatographic immunoassay (RFCI) method using gold nanoparticles as a chromatic agent. Hunter et al. [35] synthesized nanobiosensors to quickly detect pathogenic bacteria in serum with the optofluidic (surface-enhanced Raman scattering) SERS method. In this study, a hollow core photonic crystal fiber filled with microfluidics was used as the main converter. This system can be repeatedly renewed by washing with a liquid that can dissolve the analyte. The use of silver nanoparticles greatly strengthened the measuring ability of the system.

It can be said that the use of silver and gold nanoparticles in synthesized nanosensors gives very positive results in parameters such as measurement performance, sensitivity and selectivity.

Molecularly imprinted polymer receptor fluorescent sensors have several disadvantages, although they have many advantages in biology, environmental chemistry, food technology, food packaging, microbiology, pharmacology and medicine. For example, it is, unfortunately, possible to mention environmental harm properties for quantum dots in the presence of heavy metals.

3.3 Biosensors

Biosensors are formed as a result of combining the receptor and transformer components. Receptors have a biomolecular structure. Physical signals are measured by combining selectively interacting analytes, biological sensors (enzymes, antibodies, immuno-agents, nucleic acids, microorganisms, cells, tissues), and physicochemical transducers (electrodes, transistors, thermistors, optical fibers, piezoelectric crystals). Electrical signals are obtained from these measured physical signals [36]. The working principles of biosensors are based on this basis. **Figure 11** schematizes biosensors. **Table 3** summarizes the literature studies.

There are some crucial advantages of using nanotechnology in the construction of biosensors. Nanoelectronic particles increase the memory and processing capabilities of biosensors and facilitate analysis. Besides, it facilitates the identification of microorganisms, provides high selectivity and long life. It also offers the possibility to work without damaging living cells [36]. **Table 3** shows the use of some biosensors according to the literature. **Figure 12** shows some examples of biosensor technology.

Liang et al. [37] synthesized biosensors for the detection of hydrogen peroxide using carbonization with graphene nanoplates derived from ficus fruits. As is known, hydrogen peroxide is an intermediate commonly used in biological test steps, clinical diagnoses, enzymatic and many other chemical reaction environments. Since it is used so often, it is of great importance to quantitatively analyze hydrogen peroxide. It is usually analyzed by colorimetric, photochemical and electrochemical techniques. The most advantageous technique among these techniques is the electrochemical technique. Because electrochemical techniques allow fast, cheap and real-time measurements. The carbon sources are commonly used as active substances in many kinds of research.

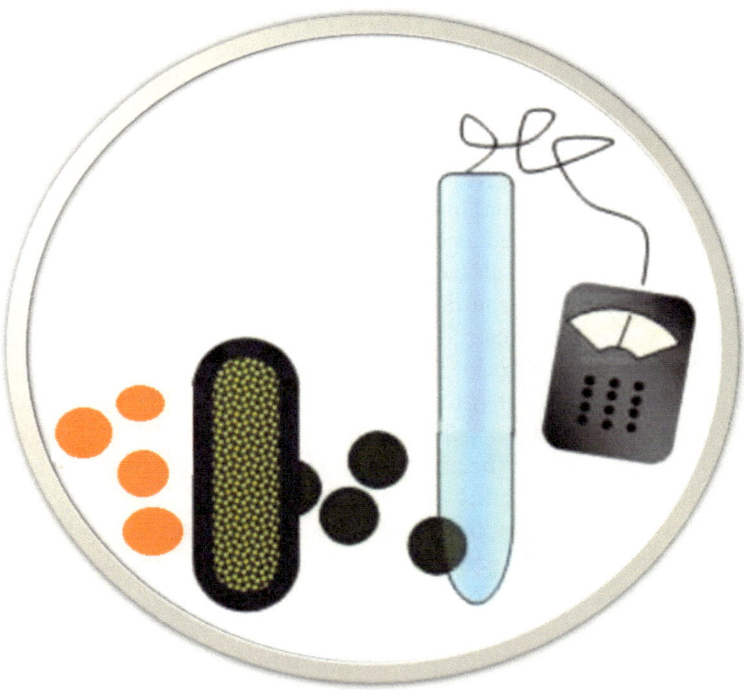

Figure 11.
Schematic representation of biosensors.

Purpose of usage	Biosensors	References
For the detection of single-nucleotide variant	Biosensor based on triplex DNA-templated Ag/Pt nanoclusters	[38]
Green synthesis of porous graphene-like nanosheets	High-sensitivity nonenzymatic hydrogen peroxide biosensor	[37]
Detection of single nucleotide polymorphism of tumor	Gold probe with lateral flow strip biosensor	[39]
For simultaneous imaging of P53 and P21 mRNA in living cells	An RGONS-based biosensor	[40]
Detection of lysozyme	Aptamer-based electrochemical biosensors	[41]
Detection of arsenic and mercury	Metallothionein-based biosensor	[42]
Detection of malachite green	Free microcantilever based biosensor	[43]
Refractive index sensor	A single-layer guided-mode resonant optical biosensor	[44]
The determination of 3-methylquinoxaline-2-carboxylic acid	Surface plasmon resonance biosensor	[45]
Detection of escherichia coli	Label-free amperometric biosensor	[46]
Electrochemical detection of the human cancer biomarker	Autocrine motility factor-phosphoglucose isomerase' based on a biosensor	[47]
Naked-eye detection of aflatoxin b1	Label-free colorimetric biosensor	[48]

Table 3.
Some examples of biosensors according to the literature.

Figure 12.
Some examples of biosensor technology.

Ma et al. [39] synthesized gold probe-based strip biosensors to diagnose liver cancer tumors. Liver cancer is the most common type of cancer in the world and is highly affected by environmental and genetic factors. Early diagnosis is vital because in some cases the patient may lose his liver only a few months after being diagnosed. In this literature study, single nucleotide polymorphism (SNP) was used as a genetic marker for gene diagnosis of the disease. Thanks to the colloidal gold strip nanoparticles, the SNP diagnosis makes exceptionally accurate diagnoses in just a few minutes. Colloidal gold strips are of great interest in the environment, food safety and medicine as they offer a fast, precise, high selectivity and inexpensive method.

Fana et al. [40] also synthesized reduced graphene oxide-based biosensors to diagnose liver cancer. Graphene oxides are preferred as nanoparticles due to their electronic, mechanical and thermal stability. However, in-vitro assays should be kept in limited amounts for use in living cells due to their toxic effects.

Khan et al. [41] synthesized aptamer-based electrochemical biosensors for lysozyme determination. These biosensors are electronic and disposable biosensors. They used the inkjet-printing method for the detection of lysozyme which is a biomarker in the diagnosis of diseases. Carbon nanotubes and the single-stranded DNA were used for aptamer immobilization on the electrode. Thus, inks containing a mixture of carbon nanotube-aptamer complexes were synthesized. Generally, the main reasons for the use of aptamers in biosensor syntheses are resistance to environmental conditions, thermal and chemical stability, and increasing the binding efficiency. In addition to these advantages, it is possible to synthesize inexpensive biosensors, have a long shelf life and can be reproduced.

Zhong et al. [49] synthesized fluorescence biosensor to detect *Pseudomonas aeruginosa* bacteria in food products. They performed the synthesis step using the copolymer points of dual-aptamer-labeled polydopamine-polyethyleneimine. According to the study of this article, it is seen that dual-aptamer biosensors enable more sensitive and accurate measurements than single-aptamer biosensors. Therefore, it has been concluded that the use of dual-aptamer-labeled polydopamine-polyethyleneimine copolymer points can be used in different alternative methods. Zhang et al. [50] synthesized colorimetric sensors for the detection of *Escherichia coli* and *Staphylococcus aureus* bacteria. It is based on the principle of separating and detecting bacteria from the medium using aptamer-based magnetic beads. Quantitative measurements of the growth kinetics of bacteria were measured by measuring the conductivity changes occurring in the environment depending on time. However, it was emphasized that the synthesized biosensor should be supported with different methods in sensitivity measurements. For this, methods such as changes in analyte volume and prolongation of incubation are recommended. Li et al. [51] used multiple amplification reactions and electrochemical methods to detect *E. coli* bacteria. First, the target sequences extracted from *E. coli* O157: H7 were converted to executive DNA and amplified. Next, a large number of transformed nucleic acid sequences were amplified by the RCA reaction. Then, DNA sequences were immobilized and electrochemical signals were measured with the help of electrochemical indicators. As a result, it is suggested that more effective results can be obtained in detecting pathogenic bacteria in living organism by developing multiple amplification methods.

Zhan et al. [52], synthesized aptasensors with the amplification method for the colorimetric detection of *Listeria monocytogenes*. In this study, enzyme dependent aptasensors were developed by rotary circle amplification (ELARCA) analysis. The study is based on the selectivity race between an aptamer specific for the bacteria of the *Listeria monocytogenes* and the biotin probe and the RCA probe. Adding bacteria to the environment prepares the medium suitable for the RCA probe, which starts the RCA process (rolling circle amplification), and causes the biotin probe to be exposed. In the presence of RCA buffer, multiple DNA copies are formed by binding to the biotin probe. In the presence of the enzyme substrate in the medium, (horse radish peroxidase), the chromophore group produced by HRP enables colorimetric measurement analysis.

As a result, it is thought that the determination of more specific aptamers for bacteria may be more developer for measurement accuracy and accuracy in the study that allows successful measurements.

4. Conclusion

Nanosystems and nanoparticles are based on the foundations of quantum physics that upset the laws of classical physics. Quantum mechanics makes it possible that nano-sized particles can be given unique and extraordinary abilities. This new technology is pushing dreams together and even promises to go beyond the borders of a futuristic imagination. Nanotechnology, which opens the door to extraordinary innovations in many engineering applications and medicine, brings many definitions such as nanoparticles, quantum dots, nanosensors, biosensors, nanospheres and nanorobots to our lives. The toxicities of quantum dots are very important for in-vivo experiments in agricultural applications. The best way to solve this problem is to secure it by bioconjugating it with coatings, proteins and peptides to protect and stabilize the surface of quantum dots. In addition, the choice of rounded quantum dots called colloidal in biosensing solutions can prevent possible problems. Using a non-toxic titanium dioxide compound can also solve the toxicity problem in in vitro applications.

Biological recognition systems connected to a transducer are effective in the specificity and selectivity of biosensors. Therefore, the most important part in a biosensor mechanism is bioreceptor synthesis. In the literature searches, biosensors for the detection of many diseases, viruses and bacteria were synthesized. However, in the applied methods, dual-aptamer biosensors generally give more precise and accurate measurements than single-aptamers. Multiple amplification reactions with fluorescence, colorimetric or electrochemical methods give better results. The more specific aptamers are used, the greater the measurement accuracy. The use of silver and gold nanoparticles in biosensor synthesis greatly increases the measuring ability of the system. Therefore, the repeatability properties of the probes will also be supported. In addition, metal organic framework compounds greatly increase the measuring ability due to their large surface areas, especially in the modification of electrodes for biosensor synthesis based on the measurement of electrochemical signals.

Even though this new technology enables rapid, inexpensive, reliable, reproducible, high-precision measurement, diagnosis and analysis in many scientific fields, the studies to be carried out in this field in the coming years will provide meaningful grounds for solving existing problems or developing more advanced technologies. It should be noted that nanotechnological systems have a significant effect on polymers. Because polymers enable easy sterilization of synthesized nanoparticles. It also increases the loading capacity of the active substance and allows the synthesis of non-toxic particles, which can be degraded and decomposed in a physiological environment. Thanks to these advantages, it supports controlled release systems and bioavailability. It makes the world of the future a candidate for the dream of the future.

Author details

Gülay Baysal
Food Engineering, Engineering Faculty, Istanbul Aydin University, Istanbul, Turkey

*Address all correspondence to: gulaybaysal@aydin.edu.tr

References

[1] Niba I, Krishnakumar K, Dineshkumar B, Nair SK. An overview on nanosphere drug delivery. European Journal of Pharmaceutıcal and Medıcal Research. 2018;**5**(4):192-198

[2] Mohan Raj VJ, Chen Y. Nanoparticle—A review. Tropical Journal of Pharmaceutical Research. 2005;**5**:561-573

[3] Jung T, Kamn W, Breitenbach A, Kaiserling E, Xiano JX. Biodegradable nanoparticle for oral delivery of peptides: Is there a role for polymers to affect mucosal uptake. European Journal of Pharmaceutics and Biopharmaceutics. 2000;**50**(1):147-160. DOI: 10.1016/S0939-6411(00)00084-9

[4] Illum L. Nanoparticulate systems for nasal delivery of drugs: A real improvement over simple systems. Journal of Pharmaceutical Sciences. 2007;**96**:473-483. DOI: 10.1002/jps.20718

[5] Ahmad M, Ahmad N, Aslam M. Synthesis of iron oxide nanorods. Advanced Science Focus. 2013;**1**: 150-155. DOI: 10.1166/asfo.2013.1014

[6] Chen S, Feng J, Guo X, Hong J, Ding W. One-step wet chemistry for preparation of magnetite nanorods. Materials Letters. 2005;**59**:985-988. DOI: 10.1016/j.matlet.2004.11.043

[7] Ghosh PK. Hydrophilic polymeric nanoparticles as drug carriers. Indian Journal of Biochemistry and Biophysics. 2000;**37**:273-282

[8] Mahapatro A, Singh DK. Biodegradable nanoparticles are excellent vehicle for site directed İn-vivo delivery of drugs and vaccines. Journal of Nanobiotechnology. 2011;**9**: 1-11. DOI: 10.1186/1477-3155-9-55

[9] Rajaonarivony M, Vauthier C, Couvrraze G, Puisieux F, Couvreur P. Development of a new drug carrier made from alginate. Journal of Pharmaceutical Sciences. 1993;**82**(9): 912. DOI: 10.1002/jps.2600820909

[10] Ren P, Wang Z, Liu B, Lu Y, Jin Z, Zhang L, et al. Highly dispersible hollow nanospheres organized by ultra-small $ZnFe_2O_4$ subunits with enhanced lithium storage properties. Journal of Alloys and Compounds. 2020;**812**:152014. DOI: 10.1016/j.jallcom.2019.152014

[11] Dong H, Haixia L, Junwei B, Li T, Baibiao H, Qingfen N. High specific surface area TiO_2 nanospheres for hydrogen production and photocatalytic activity. Journal of Nanoscience and Nanotechnology. 2020;**20**(5):3217-3224

[12] Xiong Q, Teng X, Lou J, Pan G, Xia X, Chi H, et al. Design of pyrite/carbon nanospheres as high-capacity cathode for lithium-İon batteries. Journal of Energy Chemistry. 2020;**40**: 1-6. DOI: 10.1016/j.jechem.2019.02.005

[13] Cai Y, Hua Y, Yin M, Liu H, Li S, Wang F, et al. Fabrication of test strips with gold-silver nanospheres and metal–organic frameworks: A Fluorimetric method for sensing trace cysteine in Hela cells. Sensors & Actuators: B. Chemical. 2020;**302**:127198. DOI: 10.1016/j.snb.2019.127198

[14] Chimeno-Trinchet C, Fernández-González A, García Calzón JA, Díaz-García ME, Badía LR. Alkyl-capped copper oxide nanospheres and nanoprolates for sustainability: Water treatment and İmproved lubricating performance. Science and Technology of Advanced Materials. 2019;**20**(1): 657-672. DOI: 10.1080/14686996.2019.1621683

[15] Deng K, Wang H, Xiao J, Li C, Zhang S, Huang H. Polydopamine Nanospheres loaded with L-cysteine-coated cadmium sulfide quantum dots as

Photoelectrochemical signal amplifier for PSA detection. Analytica Chimica Acta. 2019;**1090**:143-150. DOI: 10.1016/j.aca.2019.09.016

[16] Zhao D, Zhang X, Huan H, Dong X, Zhu H, Li Z, et al. Facile synthesis of MoO_3 nanospheres and their application in water treatment. Materials Letters. 2019;**256**:126648. DOI: 10.1016/j.matlet.2019.126648

[17] Jeyabanua K, Devendrana P, Manikandanb A, Packiaraja R, Rameshc K, Nallamuthu N. Preparation and characterization studies of La doped Cus nanospheres by microwave İrradiation for high performance supercapacitors. Physica B: Condensed Matter. 2019;**573**:92-101. DOI: 10.1016/j.physb.2019.08.028

[18] Özada Ç. Use of quantum dots in nuclear imaging systems. Journal of Engıneer Braıns. 2016;**1**(1):1-6

[19] Durmuşoğlu EG. Metallurgical and Materials Engineer Quantum Dots. Technical Writing. 2019. pp. 31-34. Available from: https://www.metalurji.org.tr/dergi/dergi160/d 160_3134.pdf

[20] Mousumi D, Prasad GBKS, Bisen PS. Molecular Diagnostics: Promises and Possibilities. Dordrech Heidelberg London: Springer; 2010. pp. 383-392

[21] Srivastava AK, Dev A, Karmakar S. Nanosensors and nanobiosensors in food and agriculture. Environmental Chemistry Letters. 2018;**16**:161-182. DOI: 10.1007/s10311-017-0674-7

[22] Xiang Y, Camarada MB, Wen Y, Wu H, Chen J, Li M, et al. Simple voltametric analyses of ochratoxin a in food samples using highly-stable and anti-fouling black phosphorene nanosensor. Electrochimica Acta. 2018;**282**:490-498. DOI: 10.1016/j.electacta.2018.06.055

[23] Soleja N, Manzoor O, Khan P, Mohsin M. Engineering genetically encoded FRET-based nanosensors for real time display of arsenic (As^{3+}) dynamics in living cells. Scientific Reports. 2019;**9**:11240. DOI: 10.1038/S41598-019-47682-8

[24] Dinarvand M, Neubert E, Meyer D, Selvaggio G, Mann FA, Erpenbeck L, et al. Near-infrared imaging of serotonin release from cells with fluorescent nanosensors. Nano Letters. 2019;**19**(9):6604-6611. DOI: 10.1021/acs.nanolett.9b02865

[25] Smith AF, Zhao B, You M, Jiménez JM, et al. Microfluidic DNA-based potassium nanosensors for improved dialysis treatment. Biomedical Engineering Online. 2019;**18**:73. DOI: 10.1186/S12938-019-0692-8

[26] Rathinavel S, Vadivel S, Balaji G. Development of ethanol and acetone gas sensing performance of $MgCO_2O_4$ nanosensors by clad modified fiber optical method. Optical Fiber Technology. 2019;**48**:218-224. DOI: 10.1016/j.yofte.2019.01.016

[27] Özkan A, Atar N, Yola ML. Enhanced surface plasmon resonance (SPR) signals based on İmmobilization of core-shell nanoparticles incorporated boron nitride nanosheets: Development of molecularly imprinted SPR nanosensor for anticancer drug, etoposide. Biosensors and Bioelectronics. 2019;**130**:293-298. DOI: 10.1016/j.bios.2019.01.053

[28] Satapathi S, Kumar V, Kumar Chini M, Bera R, Kanta Halder K, Patra A. Highly sensitive detection and removal of mercury İon using a multimodal. Nanosensor, Nano-Structures & Nano-Objects. 2018;**16**:120-126. DOI: 10.1016/j.nanoso.2018.05.006

[29] Rezende JDP, Max Dias Ferreira G, Max Dias Ferreira G, Mendes Da

Silva LH, Hepanhol Da Silva MDC, Soares Pinto M, et al. Polydiacetylene/triblock copolymer nanosensor for the detection of native and free bovine serum albümin. Materials Science and Engineering C. 2017;**70**:535-543. DOI: 10.1016/j.msec.2016.09.009

[30] Li Y, Zhang Z, Tao Z, Gao X, WangS LYA. Asp/Ce nanotube-based colorimetric Nanosensor for H_2O_2-free and enzyme-free detection of cysteine. Talanta. 2019;**196**:556-562. DOI: 10.1016/j.talanta.2019.01.011

[31] Qian J, Wang K, Wang C, Ren C, Liu Q, Hao N, et al. Ratiometric fluorescence nanosensor for selective and visualdetection of cadmium ions using quencher displacement-inducedfluorescence recovery of CdTe quantum dots-based hybrid probe. Sensors and Actuators B. 2017;**241**:1153-1160. DOI: 10.1016/j.snb.2016.10.020

[32] Zhou Y, Ding L, Wu Y, Huang X, Lai W, Xiong Y. Emerging strategies to develop sensitive Aunp-based ICTS nanosensors. Trends in Analytical Chemistry. 2019;**112**:147-160. DOI: 10.1016/j.trac.2019.01.006

[33] Gong X, Wang H, Liu Y, Hu Q, Gao Y, Yang Z, et al. A Di-functional and label-free carbon-based chem-nanosensor for real-time monitoring of Ph fluctuation and quantitative determining of curcumin. Analytica Chimica Acta. 2019;**1057**:132-144. DOI: 10.1016/j.aca.2019.01.012

[34] Luo K, Ryu J, Seol I, Jeong K, You SM, Kim YR. Paper-based radial chromatographic immunoassay for the detection of pathogenic bacteria in milk. Applied Materials & Interfaces. 2019;**11**(50):46472-46478. DOI: 10.1021/acsami.9b16075

[35] Hunter R, Sohi AN, Khatoon Z, Berthiaume VR, Alarcon EI, Godin M, et al. Optofluidic label-free SERS platform for rapid bacteria detection in serum. Sensors & Actuators: B. Chemical. 2019;**300**:126907. DOI: https://doi.org/10.1016/j.snb.2019.126907

[36] Tüylek Z. Biosensors and Nanotechnological interaction. BEU Journal of Science. 2017;**6**(2):71-80

[37] Liang T, Guo X, Wang J, Wei Y, Zhang D, Kong S. Green synthesis of porous graphene-like nanosheets for high-sensitivity nonenzymatic hydrogen peroxide biosensor. Materials Letters. 2019;**254**:28-32. DOI: 10.1016/j.matlet.2019.06.100

[38] Wu F, Lin Q, Wang L, Zou Y, Chen M, Xia Y, et al. A DNA electrochemical biosensor based on triplex DNA-templated Ag/Pt nanoclusters for the detection of single-nucleotide variant. Talanta. 2020;**207**:120257. DOI: 10.1016/j.talanta.2019.120257

[39] Ma LH, Wang HB, Zhang T, Xuan Y, Lia C, Chena W, et al. Visual simultaneous detection of single nucleotide polymorphism of tumor susceptibility gene and Marker alpha-fetoprotein based on double-labeled colloidal gold probe with lateral flow strip biosensor. Sensors & Actuators: B. Chemical. 2019;**298**:126819. DOI: 10.1016/j.snb.2019.126819

[40] Fana J, Tonga C, Danga W, Qin Y, Liu X, Liu B, et al. An Rgons-based biosensor for simultaneous imaging of P53 and P21 Mrna in living cells. Talanta. 2019;**204**:20-28. DOI: 10.1016/j.talanta.2019.05.087

[41] Khan NI, Maddaus AG, Song E. A low-cost inkjet-printed aptamer-based electrochemical biosensor for the selective detection of lysozyme. Biosensors. 2018;**8**:7. DOI: 10.3390/bios8010007

[42] Irvine GW, Tan SN, Stillman MJ. A simple metallothionein-based biosensor

for enhanced detection of arsenic and mercury. Biosensors. 2017;7:14. DOI: 10.3390/bios7010014

[43] Zhao Y, Gosai A, Shrotriya P. Effect of receptor attachment on sensitivity of label free microcantilever based biosensor using malachite green aptamer. Sensors & Actuators: B. Chemical. 2019;**300**:126963. DOI: 10.1016/j.snb.2019.126963

[44] Qian L, Wang K, Zhu W, Han C, Yan C. Enhanced sensing ability in a single-layer guided-mode resonant optical biosensor with deep grating. Optics Communications. 2019;**452**: 273-280. DOI: 10.1016/j. optcom.2019.07.047

[45] Peng D, Kavanagh O, Gao H, Zhang X, Deng S, Chen D, et al. Surface plasmon resonance biosensor for the determination of 3-methylquinoxaline-2-carboxylic acid, the marker residue of olaquindox, in swine tissues. Food Chemistry. 2020;**302**:124623. DOI: 10.1016/j.foodchem.2019.04.022

[46] Dhull N, Kaur G, Jain P, Mishra P, Singh D, Ganju L, et al. Label-free amperometric biosensor for *Escherichia coli* O157:H7 detection. Applied Surface Science. 2019;**495**:143548. DOI: 10.1016/j.apsusc.2019.143548

[47] Ahmad L, Salmon L, Korri-Youssoufi H. Electrochemical detection of the human cancer biomarker 'autocrine motility factor-phosphoglucose isomerase' based on a biosensor formed with a monosaccharidic inhibitör. Sensors & Actuators: B. Chemical. 2019;**299**: 126933. DOI: 10.1016/j.snb.2019.126933

[48] Wu J, Zeng L, Li N, Liu C, Chen J. A wash-free and label-free colorimetric biosensor for naked-eye detection of Aflatoxin B1 using G-Quadruplex as the signal reporter. Food Chemistry. 2019; **298**:125034. DOI: 10.1016/j. foodchem.2019.125034

[49] Zhong Z, Gao R, Chen Q, Jia L. Dual-aptamers labeled polydopamine-polyethyleneimine copolymer dots assisted engineering a fluorescence biosensor for sensitive detection of Pseudomonas aeruginosa in food samples. Spectrochimica Acta Part A: Molecular and Biomolecular Spectroscopy. 2020;**224**:117417. DOI: doi.org/10.1016/j.saa.2019.117417

[50] Zhang X, Wang X, Yang Q, Jiang X, Li Y, Zhao J, et al. Conductometric sensor for viable *Escherichia coli* and *Staphylococcus aureus* based on magnetic analyte separation via aptamer. Microchimica Acta. 2020;**187**:43. DOI: https://doi.org/10.1007/s00604-019-3880-0

[51] Li Y, Liu H, Huang H, Deng J, Fang L, Luo J, et al. A sensitive electrochemical strategy *via* multiple amplification reactions for the detection of *E. coli* O157: H7. Biosensors and Bioelectronics. 2020;**147**:111752. DOI: https://doi.org/10.1016/j. bios.2019.111752

[52] Zhan Z, Li H, Liu J, Xie G, Fang X, Wu X, et al. A competitive enzyme linked aptasensor with rolling circle amplification (ELARCA) assay for colorimetric detection of listeria monocytogenes. Food Control. 2020; **107**:106806. DOI: 10.1016/j. foodcont.2019.106806

Chapter 8

Metallic Nanowire Percolating Network: From Main Properties to Applications

Daniel Bellet, Dorina T. Papanastasiou, Joao Resende, Viet Huong Nguyen, Carmen Jiménez, Ngoc Duy Nguyen and David Muñoz-Rojas

Abstract

There has been lately a growing interest into flexible, efficient and low-cost transparent electrodes which can be integrated for many applications. This includes several applications related to energy technologies (photovoltaics, lighting, supercapacitor, electrochromism, etc.) or displays (touch screens, transparent heaters, etc.) as well as Internet of Things (IoT) linked with renewable energy and autonomous devices. This associated industrial demand for low-cost and flexible industrial devices is rapidly increasing, creating a need for a new generation of transparent electrodes (TEs). Indium tin oxide has so far dominated the field of TE, but indium's scarcity and brittleness have prompted a search into alternatives. Metallic nanowire (MNW) networks appear to be one of the most promising emerging TEs. Randomly deposited MNW networks, for instance, can present sheet resistance values below 10 Ω/sq., optical transparency of 90% and high mechanical stability under bending tests. AgNW or CuNW networks are destined to address a large variety of emerging applications. The main properties of MNW networks, their stability and their integration in energy devices are discussed in this contribution.

Keywords: transparent electrode, silver nanowire, copper nanowire, transparent conductive material, stability, percolating network

1. Introduction

Transparent electrodes (TEs) are key components for many industrial devices. TEs indeed do concern applications related to energy field such as photovoltaics or efficient lighting (light emitting diode, LED, or organic-LED, OLED), smart windows or supercapacitors and are therefore associated to rapidly increasing industrial needs. For photovoltaics, the need of TEs concerns, for instance, the front electrode that should be transparent for the sunlight while collecting the photo-generated carriers. For efficient lighting, this is the opposite physical phenomenon: injecting carriers (electrons and holes) by applying a voltage through transparent electrodes to let the generated light exit the LED or OLED device. But TEs are also used in many other applications such as transparent heaters, touch screens, sensors or radio-frequency (RF) devices.

The main TEs investigated in the last decades have been transparent conductive oxides (TCO) [1–4] with the most well-known and used one in the industrial area being indium tin oxide (ITO). And aluminium-doped zinc oxide (AZO) [2] and fluorine-doped tin oxide (FTO) [5] have been also the subject of many studies. While TCO can exhibit good or even very good physical properties, the recent industrial needs have prompted a search of new materials to replace TCO for several applications [6]. Indeed indium, for instance, can be scarce, its deposition often requires vacuum, and TCO by nature are brittle and therefore not compatible with flexible applications. Materials such as carbon nanotubes [7], graphene [8], conducting polymers [9, 10], metallic grids [11] and metallic nanowire networks [12, 13] have been mainly studied for this purpose, and some of them exhibit already promising properties for several applications. In particular, several studies have lately demonstrated that metallic nanowire (MNW) percolating networks can exhibit high electrical conductivity, high optical transparency and high flexibility [12, 14, 15]. The main investigated are silver nanowire (AgNW) and copper nanowire (CuNW). The very high aspect ratio of the nanowires (i.e. length divided by the diameter) allows these networks to achieve very good performances, similar to ITO, however by using much less raw material [12]. Such quantity are often expressed in terms of the so-called areal mass density (*amd*), defined as the required mass of metal (for MNW networks) or indium (for ITO thin layers) per square metre. Their ranges are between 40 and 200 $mg.m^{-2}$ for AgNW or CuNW networks and roughly 750–1050 $mg.m^{-2}$ for ITO thin layers [12]. With rather similar price per unit mass for both In and Ag, replacing ITO by AgNW networks appears to be a cost-effective alternative. Moreover MNW-based TEs exhibit two additional assets: they can be fabricated via solution-based methods, and they present outstanding flexibility (and even good stretchability). These two assets constitute clearly key points for an efficient industrial integration. Another advantage of MNW networks is their high optical transparency in the near-infrared spectrum, especially when compared with TCO: this is of importance for transparent solar cell applications. For those reasons, printed AgNW network-based electrodes have shown a potential as transparent and flexible electrodes in many displays such as solar cells [16–19], OLEDs [20], displays [21], supercapacitors [22], transparent heaters [23–25], radio-frequency antennas [26], antibacterial films [27] or smart windows [28].

In this contribution, we focus on TEs made of AgNWs or CuNWs and will first briefly discuss the role of the nanowire dimensions (both length and diameter) and network density on the physical properties. The network stability will be discussed followed by methods to enhance it, which appears to be a crucial issue for an efficient integration of this technology. Finally, we will briefly discuss the integration of MNW network-based transparent electrodes for energy applications.

2. Main properties of metallic nanowire networks

The main physical properties of metallic nanowire networks are optical properties (transparency and haziness), electrical resistance and mechanical properties (or more precisely the electromechanical properties mainly in bending or stretching modes). These properties depend on several parameters, including MNW dimensions (diameter and length), junction resistance and network density. We will briefly describe this dependence below. **Figure 1** shows single MNW and MNW network observed by electron microscopy at different scales.

MNW dimensions can influence the properties of MNW networks. MNW diameter, D_{NW}, can be first compared with the mean free path of electrons, Λ, in

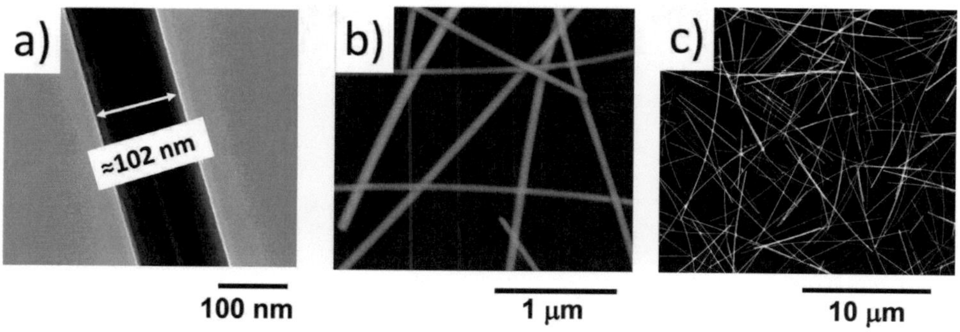

Figure 1.
Electronic microscopy observation of silver nanowires and the associated network. (a) Transmission electron microscopy image of AgNW; (b, c) scanning electron microscopy images of random AgNW network fabricated by spray on glass substrates observed at two different magnifications.

the bulk metal: if D_{NW} is comparable or even smaller than Λ, then surface scattering is increased (for bulk Ag Λ is close to 50 nm). This was derived and observed experimentally by Bid et al. at the individual MNW level [29] and shown to be in good agreement with experiments on AgNW networks by Lagrange et al. [30]. Too small MNW diameter leads to large electrical resistance and to instability at lower temperature [30], while too large MNW diameter increases shadowing effects and then reduces the optical transparency; therefore a trade-off should be found. Also one should keep in mind that large D_{NW} values lead to larger haziness. The influence of MNW length was, for instance, studied by Bergin et al. [31] or by Marus et al. [32]: generally speaking increasing MNW length results in an improvement of their optoelectronic performance. It is also worth noticing that the MNW length distribution can also play a role, as shown by Langley et al. [33], who showed that the critical density of MNW associated to the percolation threshold decreases when the MNW distribution is increased.

The junctions between MNW play also a key role. Recently Ponzoni showed that the relative contribution to electrical conductivity between nanowires and junctions could be very close [34], in good agreement with Bellew et al. who reported junction resistance measurements of individual silver nanowire junctions [35]. Bellew et al. were able to demonstrate, based on experimental data and modelling, that the junction contribution to the network's overall resistance could be reduced even beyond that of the nanowires themselves. It was shown experimentally by several methods that junctions' resistance could be reduced: for instance, Langley et al. showed that a thermal annealing can drastically reduce network resistance thanks to a local sintering of the junctions [36]; and Garnett et al. used light-induced plasmonic nano-welding to *optimize* junction resistance of MNW networks thanks to an efficient *localized* heating compatible with low-thermal-budget substrates such as polymeric substrates [37].

The network density is a key parameter and influences both the optical transmittance and the electrical resistance. Instead of considering the network density (expressed as the number of MNW per unit area), one often prefers to consider the areal mass density, amd, expressed in mass per unit area (mg/m^2). Optical transmittance is observed to decrease linearly with amd as shown by Bergin et al. [31] or by Lagrange et al. [30]. This can be simply explained by shadowing effects [30]. Conversely electrical resistance drastically decreases when amd is increased; therefore an inherent trade-off between high transparency (observed for low amd values) and low resistance (large amd values) has to be considered. **Figure 2a** illustrates the influence of amd value on the electrical resistance. Experiments performed on AgNW

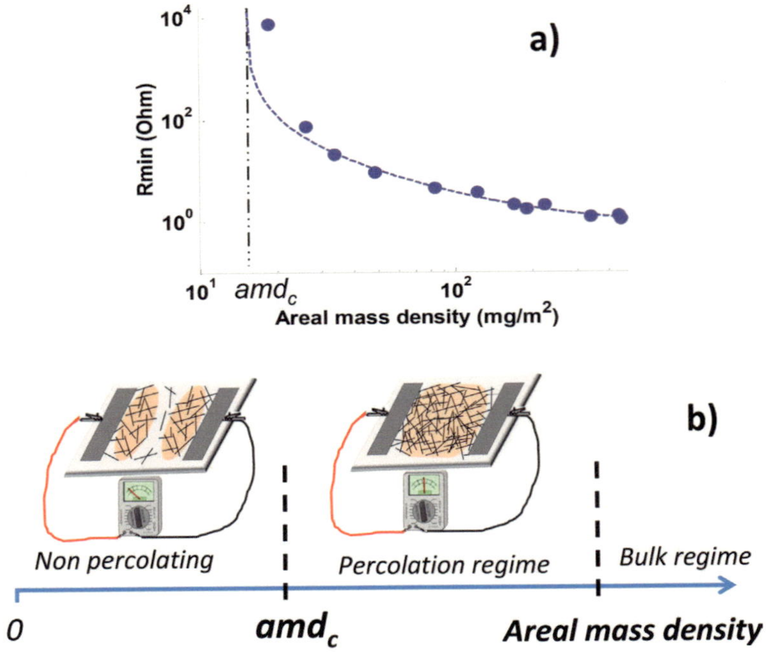

Figure 2.
Effect of the network density, expressed here as the areal mass density (amd), on the electrical properties of percolating AgNW networks: (a) Minimum electrical resistance measured experimentally during thermal ramp versus the network areal mass density [36]; (b) illustration of the three regimes associated with different values of amd. Below a critical value, amdc, no finite electrical resistance can be measured. Just above very large resistance values are measured, while they decrease as a power law for larger amd values; the onset corresponds to the stick percolation. Another transition, less known and studied, exists between the percolative regime and the bulk regime; for the latter the electrical resistivity does not depend anymore on the network density [24].

networks associated to different networks amd values show the existence of a critical value of amd, amdc, below which no finite resistance can be measured (see **Figure 2b**). This limit is associated to the stick percolation, and Monte Carlo simulations show that the amdc value is given by amdc = 5.64 < MMNW>/L2 where <MMNW > is the average mass of the MNW and L is the MNW length [33, 38]. Above amdc the measured electrical resistance is decreasing rapidly following a power law, as shown by **Figure 2a**: there is a rather good agreement observed between experimental data (symbols) and percolation theory (line) [30]. Some differences between real-world networks and Monte Carlo simulations were investigated lately by Langley et al. [33]. The real-world imperfections of a network concern the MNW length distribution, the non-isotropic MNW orientation and the MNW curvature: the influence of these three parameters on the onset network percolation was studied by Langley et al. [33]. For much larger amd value, another transition does exist between the percolative regime and the so-called bulk regime [24]; while this transition is much less known or investigated than stick percolation, such a transition occurs close to amd values that are considered in most applications.

Another way of looking at the influence of amd value on electrical properties of MNW networks is proposed in **Figure 3** where the electrical conductance is plotted versus relative amd. Below the critical amd value, amdc, the experimental resistance is infinite. Just above Sannicolo et al. demonstrated that a discontinuous activation of efficient percolating pathways takes place [39]: experimentally, for sparse networks abrupt drops of electrical resistance are observed. Such an original phenomenon was called 'geometrically quantized percolation' and was observed by lock-in thermography which evidenced the existence of individual hotter pathways

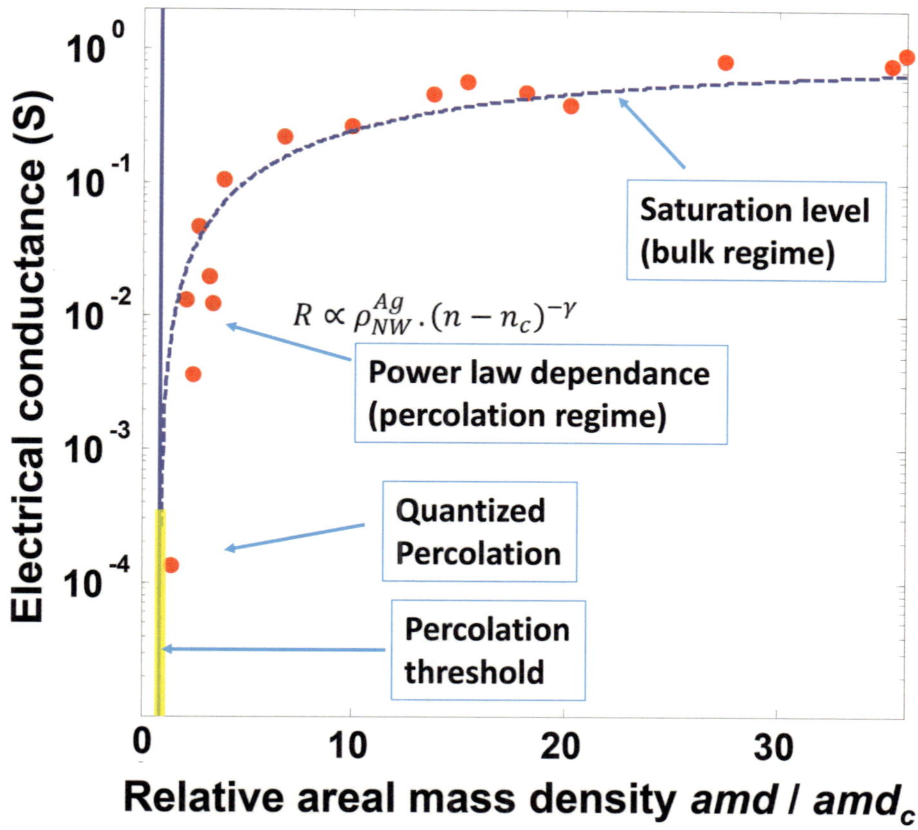

Figure 3.
Electrical conductance of AgNW network versus the relative areal mass density of the network; this shows the different electrical regimes (see **Figure 2** *and the text for more details). The red symbols correspond to experimental data, while the blue dash line corresponds to the percolation theory.*

through the network [39]. For larger amd values, one observes the percolative regime for which the electrical resistance is proportional to (amd-amdc)-γ where γ = 4/3, as shown by Lagrange et al. [30]. The previous expression has been used to fit the data of **Figure 3**, and a good agreement is observed for a large range of amd values, while percolation theory should be only valid for amd values slightly above amdc. For very large values of amd, the electrical resistivity does not depend upon amd value, and a metallic bulk behaviour should be then observed.

3. Stability of silver nanowire networks

The stability of metallic nanowire networks appears as a crucial issue, specifically when such TEs undergo thermal and electrical stress [29, 30]. This concerns nearly all applications, and the stability is related to electrical and thermal stability but also long-term ageing and chemical degradation. Such instability can stem from different physical mechanisms such as diffusion of metallic atoms, electromigration processes during electrical stress or oxidation of silver or copper if networks are in contact with either humid atmosphere and/or in high temperature conditions and/or under electrical stress.

One of the first investigations of the instability of AgNW networks was reported by Khaligh and Goldthorpe [40]: they showed that when AgNW-based TEs undergo similar electrical currents than those encountered in organic solar cells, the TEs

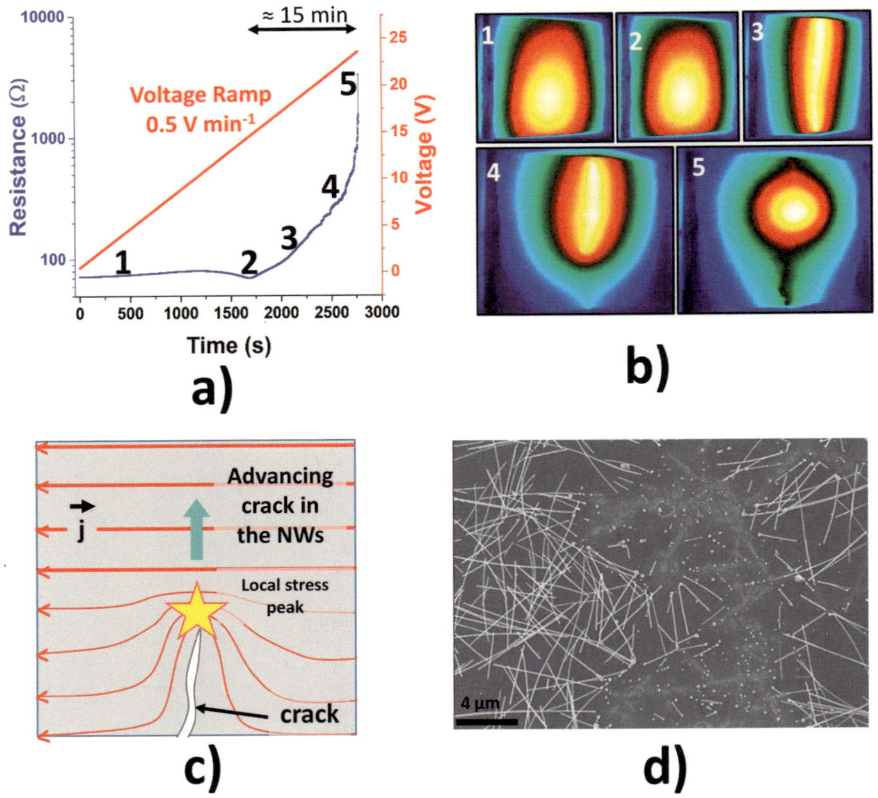

Figure 4.
Electrical failure observations. (a) Time dependence of the electrical resistance of a AgNW network during an electrical ramp of 0.5 V min^{-1}; the electrical breakdown is observed for voltage larger than 9 volts. (b) Corresponding thermal maps captured in situ with an IR camera at specific times during the experiment depicted in (a) (the electrodes at opposite sides of the specimen are vertical); (c) schematic representation of the mechanisms involved in the crack explaining the crack propagation; (d) scanning electron microscopy observation of the AgNW network location where the crack took place during electrical breakdown.

can quickly fail within 2 days. They reported that such failure is associated with local Joule heating which deteriorates AgNW network and eventually leads to the network failure. Similar observations were also reported by Chen et al. [41]. Such stability issues are even more pronounced for CuNWs since oxidation (even at low temperature) of CuNWs can occur [42–44].

Figure 4 exhibits the electrical failure observations. **Figure 4a** shows the typical time dependence of the AgNW network electrical resistance during an electrical ramp with the electrical breakdown observed for voltage larger than 9 volts. At first a slight increase of the resistance is observed: thanks to the Joule effect, the network temperature is slightly increased, and since its behaviour corresponds to a metal, its electrical resistance thus increases. For voltage larger than 9 volts, the network electrical resistance drastically increases: this is associated to the electrical breakdown of the network [45].

Figure 4b exhibits in situ thermal maps of the same specimen considered in **Figure 4a** with identical corresponding numbers during the voltage ramp. During the degradation phase (i.e. between (2) and (4)), the heat distribution appears to narrow to a vertical central part of the network parallel to the contact electrodes. At step 4, the accelerated increase in resistance can be associated to the occurrence of a 'thermal' crack which is clearly detectable at the bottom at step 5. A schematic representation of the involved mechanism is shown **Figure 4c**: the propagation mechanism of the crack is related to the displacement of the local current stress peak, which keeps being constricted to the top extremity of the deteriorated area, leading to a runaway-like destruction phenomenon. Finally as shown by **Figure 4d**, AgNWs located close to the crack do appear fully or partially spheroidized, as a result of a drastic local heating and/or electromigration.

To mitigate such instability problems, several studies showed that coating of MNW networks with either inorganic nanoparticles or thin films can drastically improve MNW network stability and integration into real devices. One can cite Morgenstern et al. [46] who showed that solution-processed AgNW films coated with ZnO nanoparticles enhance performance of such TEs when integrated in organic solar cells. The observed efficient protection against AgNW oxidation or degradation leads to a photocurrent for the organic solar cell which appears to be enhanced when compared with ITO use [46]. And, Göbelt et al. showed that a thin aluminium-doped zinc oxide layer deposited by atomic layer deposition (ALD) on AgNW leads to similar photovoltaic performances [47], however by using a much lower silver amount. Some other studies also recently reported that the coating with very thin film layers of titanium dioxide (TiO_2) [48] and zinc oxide (ZnO) [49] clearly enhances the stability. It is worth noticing that new depositing approaches have been assessed lately, and one of the most promising appears to be the atmospheric pressure spatial atomic layer deposition (AP-SALD) [50, 51]. While maintaining the advantages of ALD (viz., low-temperature deposition, thickness control, high-quality materials, and conformity), it can be much faster (up to 2 orders of magnitude faster) than ALD. Another clear asset is its compatibility with roll-to-roll and open air technology. Lately, our team used AP-SALD [52] to fabricate nanocomposite-based TEs in which AgNWs are protected by a conformal thin oxide layer (ZnO). The AP-SALD method was used to deposit thin layers of 15–30 nm ZnO around the AgNWs with the goal of enhancing the network stability [52]. The ZnO coating improved the adhesion of the AgNW networks to the glass substrate, which is known to be poor for bare AgNW networks. **Figure 5** illustrates the positive effects of a coating on MNW-based TE by showing the evolution of the electrical resistance of bare or ZNO-coated AgNW networks during a voltage ramp of 0.1 V/min. A clear enhancement of electrical stability is observed, reducing the degradation of such TEs. As shown by **Figure 5** and by Khan et al. [52], the thicker

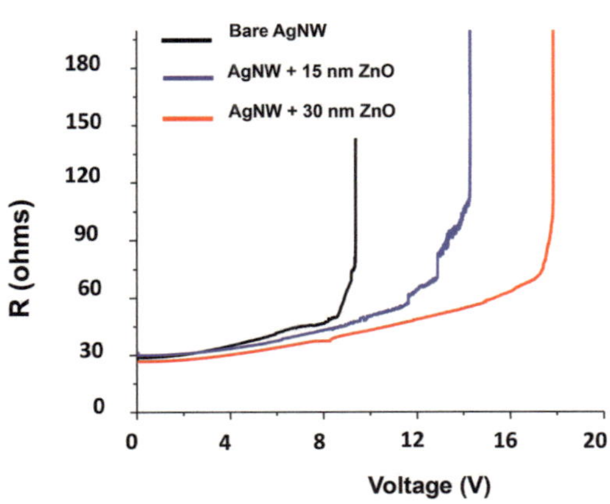

Figure 5.
Electrical failure observations and stability enhancement thanks to coating of MNWs. Variation of electrical resistance for bare and ZnO-coated AgNW networks when subjected to voltage ramps of 0.1 V/min. The bare AgNW network shows failure at around 9 V, whereas the stability of ZnO-coated AgNW networks increases with the increasing ZnO coating thickness to 14 and 18 volts for, respectively, 15 and 30 nm of ZnO coating.

the deposited ZnO layer, the better the stability. This stability enhancement can be explained as follows: the ZnO oxide coating can (at least partially) hinder silver atomic diffusion through the oxide coating [52], avoiding the spheroidization and/or electromigration of AgNWs. A compromise in terms of oxide coating thickness has to be considered depending on the target application since the thicker the ZnO coating, the lower the optical transparency.

Another example of stability enhancement was reported by Shi et al. [53] who demonstrated that transfer of CVD grown graphene onto CuNW films drastically enhances the stability of the hybrid films over long time scale (up to 180 days), while different ageing conditions were also investigated. Graphene is shown to play a key role for preventing oxygen species permeation which drastically decreases oxidation rate. This allows to obtain stable CuNW networks associated with both high optical-electrical performance and excellent stability [53].

In summary to avoid any degradation or oxidation, metallic nanowires are nowadays often coated by a protective layer for an improved integration. This protective (nanoparticles or thin inorganic) layer could be either metallic [54], based on graphene, polymeric or a transparent oxide [52]. This leads in general to a much enhanced thermal and electrical stability along with a better adhesion, although this is at the expense of optical transmittance decrease. One can also observe that conformal thin oxide coating deposited either by ALD or by spatial ALD appears to be an efficient protecting coating while keeping rather high network transmittance [52].

4. Use of metallic nanowire networks for energy applications

Among others, photovoltaic systems, light-emitting diodes (LED) or smart windows constitute sustainable green energy technologies which have been intensively studied lately for energy saving and/or for an alternative to fossil fuel energy. Generally speaking the main goals associated with these technologies concern cost reduction, efficiency improvement and use of abundant materials. For photovoltaic and efficient lighting (LED or OLED), the light should either enter a solar cell or exit the LED requiring the use of an efficient transparent electrode for, respectively, collecting or injecting the carriers. Several investigations have shown that MNW network-based transparent electrodes can be efficiently integrated in such energy devices thanks to their electrical and optical properties. Their excellent flexibility constitutes a clear asset for flexible devices and/or when fast (and then low-cost) technologies such as roll-to-roll are used for the industrial fabrication. And the possibility to coat MNWs allows to tune the work function and band alignments and can therefore lead to better integration possibilities.

For solar cells, MNW networks have been mainly tested in organic solar cells. One of the first demonstrations was reported by Leem et al. [55]: these authors used AgNW network as electrode in P3HT/PCBM organic solar cells, and it showed an efficiency of 2.5% which was equivalent to ITO-based devices. And Yang et al. showed that by using fully solution-processed polymer, bulk heterojunction (BHJ) solar cells with anodes composed of AgNWs were successfully fabricated with performances slightly lower than when ITO is used [56]. Interestingly they showed that the BHJ solar cells were highly flexible since the fabricated solar cells exhibited recoverable efficiency even under large bending deformation up to 120°.

AgNW and CuNW can also be efficiently integrated in OLED devices. AgNW networks were the first to be integrated in OLED devices: The obtained electrode was shown to be suitable for the fabrication of high-performance polymer-based LED [57]. A very recent study by Lian et al. reported the use of CuNW-based composite film for OLED integration [58]. A good electrical conductivity (22 Ω/sq), high transmittance (88%), low surface roughness and good adherence to the substrate were observed. The good adherence originates from the presence of the polymethyl methacrylate (PMMA) coating on CuNWs. The fabricated CuNW/PMMA composite film appears to be stable since it can resist air, water and ethanol exposure without electrical deterioration. With this CuNW/PMMA composite film as anode for the OLED, the device performances appear even better than those with using ITO anode [58].

Several articles clearly also showed that MNW network-based transparent electrodes can be efficiently integrated within electrochromic devices. Thanks to the chrono-amperometry curves and the corresponding in situ transmittance curve at 1100 nm for a WO_3/AgNW film, it was clearly demonstrated that good performances can be obtained when the fabricated electrochromic device uses AgNW network [59]. Such fabricated films exhibit excellent cycling stability as well as distinct modulation of near-infrared light compared with ITO-based electrochromic devices.

Another energy application which appears promising for MNW integration in energy area concerns supercapacitor. Yuksel et al. [22] fabricated nanocomposite electrochromic supercapacitor electrodes: AgNW network electrodes and a green to transmissive electrochromic polymer (PDOPEQ). These authors showed that the obtained supercapacitors have a changing colour from vibrant green to transparent with very good characteristics (i.e. specific capacitance of 61.5 F/g at a current density of 0.1 A/g, capacity retention upon 20,000 galvanostatic charge-discharge cycles). Such characteristics appear very promising for use of MNW-based transparent electrodes in electrochromic supercapacitors [22].

5. Conclusions

As discussed MNW networks exhibit strong potential to act as efficient transparent electrodes for many applications. Indeed, MNW exhibits high transparency and low electrical resistance levels, which are associated with excellent bendability and good stretchability. This contribution reports briefly the influence of main parameters on the MNW network-based TE, the prevailing parameters being MNW chemical nature and dimensions as well as network density. Still, for approaching an efficient integration into industrial devices such as organic solar cells, efficient light production or smart windows, several other requirements have to be considered. One of the most important ones concerns their stability which appears to be a crucial issue: it can involve either electrical, thermal and mechanical aspects or ageing and chemical degradation. The origin of failure in MNW networks was discussed in this contribution with the following stages: optimization, degradation and breakdown of the MNW network. The breakdown occurs via a localized mechanism thanks to the creation and propagation of a crack. To prevent such instability, encapsulation of MNW network is performed by thin oxide layers: this leads to a drastic enhancement of the MNW networks stability. Moreover such approach shows improved adhesion and much better thermal and electrical stability. Finally this contribution shows that the scientific community has worked in several directions and has demonstrated that MNW network-based transparent electrodes can be integrated in industrial devices such as organic photovoltaic, light-emitting diode, in smart windows or supercapacitors. The prospects concern the replacement of AgNW by cheaper MNW such as CuNW while the stability might be a stronger issue than for AgNW, a better optimization of the many parameters (MNW chemistry and dimensions, coating, etc.) for a given application and to make MNW deposition and optimization fast and low-cost enough to be compatible with industrial challenges (for instance, compatible with the very fast roll-to-roll technology).

Acknowledgements

The authors warmly thank Y. Bréchet, C. Celle, D.P. Langley, T. Sannicolo, S. Aghazadehchors and J.-P. Simonato for fruitful discussions. This work was performed within the framework of the Centre of Excellence of Multifunctional Architectured Materials 'CEMAM' n° AN-10-LABX-44-01. This work was funded by the Agence Nationale de Recherche (ANR, France) via the programme ANR-16-CE05-0021 (Despatch), ANR-18-CE09-0041 (Meaning), ANR-18-CE09-0036 (Panassé) and ANR-15-CE05-0019 (Indeed). This work was as well supported by Région Auvergne Rhône-Alpes through the project Pack Ambition Recherche 2018 Eternité. The Carnot Energies du Futur is acknowledged through the project FREE.

Conflict of interest

The authors declare no conflict of interest.

Author details

Daniel Bellet[1*†], Dorina T. Papanastasiou[1], Joao Resende[1], Viet Huong Nguyen[2,3], Carmen Jiménez[1], Ngoc Duy Nguyen[4] and David Muñoz-Rojas[1*†]

1 Université Grenoble Alpes, CNRS, Grenoble INP, LMGP, Grenoble, France

2 Faculty of Electrical and Electronic Engineering, Phenikaa University, Hanoi, Vietnam

3 Phenikaa Research and Technology Institute (PRATI), A&A Green Phoenix Group JSC, Hanoi, Vietnam

4 Département de Physique, Université de Liège, CESAM/Q-MAT, SPIN, Liège, Belgium

*Address all correspondence to: daniel.bellet@grenoble-inp.fr and david.munoz-rojas@grenoble-inp.fr

† These authors contributed equally to this work.

References

[1] Ellmer K. Past achievements and future challenges in the development of optically transparent electrodes. Nature Photonics. 2012;**6**:809-817. DOI: 10.1038/nphoton.2012.282

[2] Ellmer K, Klein A, Rech B. Transparent Conductive Zinc Oxide: Basics and Applications in Thin Film Solar Cells. Berlin: Springer; 2008

[3] Fortunato E, Ginley D, Hosono H, Paine DC. Transparent conducting oxides for photovoltaics. MRS Bulletin. 2007;**32**:242-247

[4] Klein A. Transparent conducting oxides: Electronic structure–property relationship from photoelectron spectroscopy with in situ sample preparation. Journal of the American Ceramic Society. 2013;**96**:331-345. DOI: 10.1111/jace.12143

[5] Giusti G, Consonni V, Puyoo E, Bellet D. High performance ZnO-SnO_2:F Nanocomposite transparent electrodes for energy applications. ACS Applied Materials and Interfaces. 2014;**6**:14096-14107. DOI: 10.1021/am5034473

[6] Hecht DS, Hu L, Irvin G. Emerging transparent electrodes based on thin films of carbon nanotubes, graphene, and metallic nanostructures. Advanced Materials. 2011;**23**:1482-1513

[7] Zhou Y, Azumi R. Carbon nanotube based transparent conductive films: Progress, challenges, and perspectives. Science and Technology of Advanced Materials. 2016;**17**:493-516. DOI: 10.1080/14686996.2016.1214526

[8] López-Naranjo EJ, González-Ortiz LJ, Apátiga LM, Rivera-Muñoz EM, Manzano-Ramírez A. Transparent electrodes: A review of the use of carbon-based Nanomaterials. Journal of Nanomaterials. 2016;**2016**:1-12. DOI: 10.1155/2016/4928365

[9] Yang J, Liu Y, Liu S, Li L, Zhang C, Liu T. Conducting polymer composites: Material synthesis and applications in electrochemical capacitive energy storage. Materials Chemistry Frontiers. 2017;**1**:251-268. DOI: 10.1039/C6QM00150E

[10] Gueye MN, Carella A, Demadrille R, Simonato J-P. All-polymeric flexible transparent heaters. ACS Applied Materials and Interfaces. 2017;**9**:27250-27256. DOI: 10.1021/acsami.7b08578

[11] Park JH, Lee DY, Kim Y-H, Kim JK, Lee JH, Park JH, et al. Flexible and transparent metallic grid electrodes prepared by evaporative assembly. ACS Applied Materials and Interfaces. 2014;**6**:12380-12387. DOI: 10.1021/am502233y

[12] Sannicolo T, Lagrange M, Cabos A, Celle C, Simonato J-P, Bellet D. Metallic nanowire-based transparent electrodes for next generation flexible devices: A review. Small. 2016;**12**:6052-6075. DOI: 10.1002/smll.201602581

[13] van de Groep J, Spinelli P, Polman A. Transparent conducting silver nanowire networks. Nano Letters. 2012;**12**:3138-3144. DOI: 10.1021/nl301045a

[14] Guo CF, Ren Z. Flexible transparent conductors based on metal nanowire networks. Materials Today. 2015;**18**:143-154. DOI: 10.1016/j.mattod.2014.08.018

[15] Bellet D, Lagrange M, Sannicolo T, Aghazadehchors S, Nguyen VH, Langley DP, et al. Transparent electrodes based on silver nanowire networks: From physical considerations towards device integration. Materials. 2017;**10**:570. DOI: 10.3390/ma10060570

[16] Langley DP, Giusti G, Lagrange M, Collins R, Jiménez C, Bréchet Y, et al. Silver nanowire networks: Physical

properties and potential integration in solar cells. Solar Energy Materials and Solar Cells. 2014;**125**:318-324. DOI: 10.1016/j.solmat.2013.09.015

[17] Guo F, Li N, Radmilović VV, Radmilović VR, Turbiez M, Spiecker E, et al. Fully printed organic tandem solar cells using solution-processed silver nanowires and opaque silver as charge collecting electrodes. Energy and Environmental Science. 2015;**8**:1690-1697. DOI: 10.1039/C5EE00184F

[18] Yu Y-Y, Ting Y-J, Chung C-L, Tsai T-W, Chen C-P. Comprehensive study on chemical and hot press-treated silver nanowires for efficient polymer solar cell application. Polymers. 2017;**9**:635. DOI: 10.3390/polym9110635

[19] Chalh M, Vedraine S, Lucas B, Ratier B. Plasmonic Ag nanowire network embedded in zinc oxide nanoparticles for inverted organic solar cells electrode. Solar Energy Materials and Solar Cells. 2016;**152**:34-41. DOI: 10.1016/j.solmat.2016.03.021

[20] Coskun S, Selen Ates E, Emrah Unalan H. Optimization of silver nanowire networks for polymer light emitting diode electrodes. Nanotechnology. 2013;**24**:125202. DOI: 10.1088/0957-4484/24/12/125202

[21] Park H-G, Heo G-S, Park S-G, Jeong H-C, Lee JH, Seo D-S. Silver nanowire networks as transparent conducting films for liquid crystal displays. ECS Solid State Letters. 2015;**4**:R50-R52. DOI: 10.1149/2.0031510ssl

[22] Yuksel R, Coskun S, Gunbas G, Cirpan A, Toppare L, Unalan HE. Silver nanowire/conducting polymer nanocomposite electrochromic supercapacitor electrodes. Journal of the Electrochemical Society. 2017;**164**:A721-A727. DOI: 10.1149/2.0791704jes

[23] Celle C, Mayousse C, Moreau E, Basti H, Carella A, Simonato J-P. Highly flexible transparent film heaters based on random networks of silver nanowires. Nano Research. 2012;**5**:427-433. DOI: 10.1007/s12274-012-0225-2

[24] Sorel S, Bellet D, Coleman JN. Relationship between material properties and transparent heater performance for both bulk-like and Percolative nanostructured networks. ACS Nano. 2014;**8**:4805-4814. DOI: 10.1021/nn500692d

[25] Ergun O, Coskun S, Yusufoglu Y, Unalan HE. High-performance, bare silver nanowire network transparent heaters. Nanotechnology. 2016;**27**:445708. DOI: 10.1088/0957-4484/27/44/445708

[26] Song L, Myers AC, Adams JJ, Zhu Y. Stretchable and reversibly deformable radio frequency antennas based on silver nanowires. ACS Applied Materials and Interfaces. 2014;**6**:4248-4253. DOI: 10.1021/am405972e

[27] Chen Y, Lan W, Wang J, Zhu R, Yang Z, Ding D, et al. Highly flexible, transparent, conductive and antibacterial films made of spin-coated silver nanowires and a protective ZnO layer. Physica E: Low-dimensional Systems and Nanostructures. 2016;**76**:88-94. DOI: 10.1016/j.physe.2015.10.009

[28] Yuksel R, Ataoglu E, Turan J, Alpugan E, Ozdemir Hacioglu S, Toppare L, et al. A new high-performance blue to transmissive electrochromic material and use of silver nanowire network electrodes as substrates. Journal of Polymer Science Part A: Polymer Chemistry. 2017;**55**:1680-1686. DOI: 10.1002/pola.28506

[29] Bid A, Bora A, Raychaudhuri A. Observation of large low-frequency resistance fluctuations in metallic nanowires: Implications on its stability.

Physical Review B. 2005;**72**. DOI: 10.1103/PhysRevB.72.113415

[30] Lagrange M, Langley DP, Giusti G, Jiménez C, Bréchet Y, Bellet D. Optimization of silver nanowire-based transparent electrodes: Effects of density, size and thermal annealing. Nanoscale. 2015;**7**:17410-17423. DOI: 10.1039/C5NR04084A

[31] Bergin SM, Chen Y-H, Rathmell AR, Charbonneau P, Li Z-Y, Wiley BJ. The effect of nanowire length and diameter on the properties of transparent, conducting nanowire films. Nanoscale. 2012;**4**:1996. DOI: 10.1039/c2nr30126a

[32] Marus M, Hubarevich A, Lim RJW, Huang H, Smirnov A, Wang H, et al. Effect of silver nanowire length in a broad range on optical and electrical properties as a transparent conductive film. Optical Materials Express. 2017;**7**:1105. DOI: 10.1364/OME.7.001105

[33] Langley DP, Lagrange M, Nguyen ND, Bellet D. Percolation in networks of 1-dimensional objects: Comparison between Monte Carlo simulations and experimental observations. Nanoscale Horizons. 2018;**3**:545-550. DOI: 10.1039/C8NH00066B

[34] Ponzoni A. The contributions of junctions and nanowires/nanotubes in conductive networks. Applied Physics Letters. 2019;**114**:153105. DOI: 10.1063/1.5090117

[35] Bellew AT, Manning HG, Gomes da Rocha C, Ferreira MS, Boland JJ. Resistance of single Ag nanowire junctions and their role in the conductivity of nanowire networks. ACS Nano. 2015;**9**:11422-11429. DOI: 10.1021/acsnano.5b05469

[36] Langley DP, Lagrange M, Giusti G, Jiménez C, Bréchet Y, Nguyen ND, et al. Metallic nanowire networks: Effects of thermal annealing on electrical resistance. Nanoscale. 2014;**6**:13535-13543. DOI: 10.1039/C4NR04151H

[37] Garnett EC, Cai W, Cha JJ, Mahmood F, Connor ST, Greyson Christoforo M, et al. Self-limited plasmonic welding of silver nanowire junctions. Nature Materials. 2012;**11**:241-249. DOI: 10.1038/nmat3238

[38] Li J, Zhang S-L. Finite-size scaling in stick percolation. Physical Review E. 2009;**80**:040104. DOI: 10.1103/PhysRevE.80.040104

[39] Sannicolo T, Muñoz-Rojas D, Nguyen ND, Moreau S, Celle C, Simonato J-P, et al. Direct imaging of the onset of electrical conduction in silver nanowire networks by infrared thermography: Evidence of geometrical quantized percolation. Nano Letters. 2016;**16**:7046-7053. DOI: 10.1021/acs.nanolett.6b03270

[40] Khaligh HH, Goldthorpe IA. Failure of silver nanowire transparent electrodes under current flow. Nanoscale Research Letters. 2013;**8**:1-6

[41] Chen D, Zhao F, Tong K, Saldanha G, Liu C, Pei Q. Mitigation of electrical failure of silver nanowires under current flow and the application for long lifetime organic light-emitting diodes. Advanced Electronic Materials. 2016;**2**:1600167. DOI: 10.1002/aelm.201600167

[42] Celle C, Cabos A, Fontecave T, Laguitton B, Benayad A, Guettaz L, et al. Oxidation of copper nanowire based transparent electrodes in ambient conditions and their stabilization by encapsulation: Application to transparent film heaters. Nanotechnology. 2018;**29**:085701. DOI: 10.1088/1361-6528/aaa48e

[43] Xu L, Yang Y, Hu Z-W, Yu S-H. Comparison study on the stability of copper nanowires and their oxidation

kinetics in gas and liquid. ACS Nano. 2016;**10**:3823-3834. DOI: 10.1021/acsnano.6b00704

[44] Mock J, Bobinger M, Bogner C, Lugli P, Becherer M, Mock J, et al. Aqueous synthesis, degradation, and encapsulation of copper nanowires for transparent electrodes. Nanomaterials. 2018;**8**:767. DOI: 10.3390/nano8100767

[45] Sannicolo T, Charvin N, Flandin L, Kraus S, Papanastasiou DT, Celle C, et al. Electrical mapping of silver nanowire networks: A versatile tool for imaging network homogeneity and degradation dynamics during failure. ACS Nano. 2018;**12**:4648-4659. DOI: 10.1021/acsnano.8b01242

[46] Morgenstern FSF, Kabra D, Massip S, Brenner TJK, Lyons PE, Coleman JN, et al. Ag-nanowire films coated with ZnO nanoparticles as a transparent electrode for solar cells. Applied Physics Letters 2011;4:183307-183307-3. DOI: 10.1063/1.3656973

[47] Göbelt M, Keding R, Schmitt SW, Hoffmann B, Jäckle S, Latzel M, et al. Encapsulation of silver nanowire networks by atomic layer deposition for indium-free transparent electrodes. Nano Energy. 2015;**16**:196-206. DOI: 10.1016/j.nanoen.2015.06.027

[48] Song T-B, Rim YS, Liu F, Bob B, Ye S, Hsieh Y-T, et al. Highly robust silver nanowire network for transparent electrode. ACS Applied Materials and Interfaces. 2015;**7**:24601-24607. DOI: 10.1021/acsami.5b06540

[49] Pham A-T, Nguyen X-Q, Tran D-H, Ngoc Phan V, Duong T-T, Nguyen D-C. Enhancement of the electrical properties of silver nanowire transparent conductive electrodes by atomic layer deposition coating with zinc oxide. Nanotechnology. 2016;**27**:335202. DOI: 10.1088/0957-4484/27/33/335202

[50] Muñoz-Rojas D, MacManus-Driscoll J. Spatial atmospheric atomic layer deposition: A new laboratory and industrial tool for low-cost photovoltaics. Materials Horizons. 2014;**1**:314-320. DOI: 10.1039/C3MH00136A

[51] Muñoz-Rojas D, Nguyen VH, Masse de la Huerta C, Aghazadehchors S, Jiménez C, Bellet D. Spatial atomic layer deposition (SALD), an emerging tool for energy materials. Application to new-generation photovoltaic devices and transparent conductive materials. Comptes Rendus Physique. 2017;**18**:391-400. DOI: 10.1016/j.crhy.2017.09.004

[52] Khan A, Nguyen VH, Muñoz-Rojas D, Aghazadehchors S, Jiménez C, Nguyen ND, et al. Stability enhancement of silver nanowire networks with conformal ZnO coatings deposited by atmospheric pressure spatial atomic layer deposition. ACS Applied Materials and Interfaces. 2018;**10**:19208-19217. DOI: 10.1021/acsami.8b03079

[53] Shi L, Wang R, Zhai H, Liu Y, Gao L, Sun J. A long-term oxidation barrier for copper nanowires: Graphene says yes. Physical Chemistry Chemical Physics. 2015;**17**:4231-4236. DOI: 10.1039/C4CP05187D

[54] Stewart IE, Ye S, Chen Z, Flowers PF, Wiley BJ. Synthesis of Cu–Ag, Cu–Au, and Cu–Pt Core–Shell nanowires and their use in transparent conducting films. Chemistry of Materials. 2015. DOI: 10.1021/acs.chemmater.5b03709

[55] Leem D-S, Edwards A, Faist M, Nelson J, Bradley DDC, de Mello JC. Efficient organic solar cells with solution-processed silver nanowire electrodes. Advanced Materials. 2011;**23**:4371-4375. DOI: 10.1002/adma.201100871

[56] Yang L, Zhang T, Zhou H, Price SC, Wiley BJ, You W. Solution-processed flexible polymer solar cells with silver nanowire electrodes. ACS Applied Materials and Interfaces. 2011;**3**:4075-4084. DOI: 10.1021/am2009585

[57] Yu Z, Zhang Q, Li L, Chen Q, Niu X, Liu J, et al. Highly flexible silver nanowire electrodes for shape-memory polymer light-emitting diodes. Advanced Materials. 2011;**23**:664-668. DOI: 10.1002/adma.201003398

[58] Lian L, Dong D, Wang H, He G. Highly reliable copper nanowire electrode with enhanced transmittance and robustness for organic light emitting diodes. Organic Electronics. 2019;**65**:70-76. DOI: 10.1016/j.orgel.2018.11.002

[59] Zhou K, Wang H, Zhang S, Jiu J, Liu J, Zhang Y, et al. Electrochromic modulation of near-infrared light by WO_3 films deposited on silver nanowire substrates. Journal of Materials Science. 2017;**52**:12783-12794. DOI: 10.1007/s10853-017-1391-0

Chapter 9

Emerging Artificial Two-Dimensional van der Waals Heterostructures for Optoelectronics

Hongcheng Ruan, Yu Huang, Yuqian Chen and Fuwei Zhuge

Abstract

Two-dimensional (2D) materials are attracting explosive attention for their intriguing potential in versatile applications, covering optoelectronics, electronics, sensors, etc. An attractive merit of 2D materials is their viable van der Waals (VdW) stacking in artificial sequence, thus forming different atomic arrangements in vertical direction and enabling unprecedented tailoring of material properties and device application. In this chapter, we summarize the latest progress in assembling VdW heterostructures for optoelectronic applications by beginning with the basic pick-transfer method for assembling 2D materials and then discussing the different combination of 2D materials of semiconductor, conductor, and insulator properties for various optoelectronic devices, e.g., photodiode, phototransistors, optical memories, etc.

Keywords: heterostructures, optoelectronics, van der Waals, 2D materials, artificial stacking

1. Introduction

Since the successful exfoliation of graphene [1], a group of materials with two-dimensional structures have revived and are attracting explosive interests from a variety of fields, including transistors [2], photodetectors [3], chemical sensors, memories, and artificial synapses [4, 5]. This is benefited from the versatile properties, of 2D materials defined not only by their crystal structure (1 T, 2H, etc.) but also by their layer number, i.e., the electrical conductivity and optical bandgaps [6]. The transition metal chalcogenides (TMDs) of 1 T or 1 T' phase usually manifest metallic behavior, while in 2H phase, they are semiconductor and can be transformed into insulator by field-effect modulation [7]. Meanwhile, monolayer MoS_2, WSe_2, and $MoTe_2$ are transformed into direct band semiconductor with greatly improved photoluminescence yield compared to their indirect bulk form, rendering the further fabrication of light emitting diodes [8, 9]. The recent appearance of 2D ferroelectric materials from direct chemical synthesis or atom doping has further enriched the physical properties of 2D semiconductors [10, 11]. These rapid evolution of 2D materials with diverse physical and chemical properties motivates enduring efforts to explore various property tuning and integration strategies in functional devices, e.g., via chemical doping, alloying, or constructing heterostructures [12].

An indispensable feature of the 2D materials is their van der Waals interlayer coupling, which is weak enough compared to covalent or ionic bonding to enable mechanical or electrochemical exfoliation [13]. The exfoliated 2D materials in monolayer or few layer thicknesses can then be artificially stacked, either laterally or vertically, making heterostructures in various forms that are not possible in conventional semiconductors with 3D crystal lattice (Si, III-V, and oxides) due to the lattice mismatch. The great flexibility in assembling 2D materials thus renders unprecedented opportunity in discovering novel nanoscale transport phenomenon [14] and carrier dynamics and stimulates the exploration of 2D functional devices via deliberately designing the heterostructures. In optoelectronics, this enabled the tailoring of charge separation characteristics of photogenerated electron–hole pairs in semiconductors [15], thereby allowing innovated designs of heterostructured transistors [16, 17], tunneling diode for photodetection [18, 19], and further optoelectronic memories with float gate structures [20].

In this chapter, we first introduce the basic design of heterostructures for optoelectronics and the pick-transfer methods for their artificial assembly and then discuss the recent progress in fabricating novel 2D vdW heterostructures for functional devices. In view of the rapid progress in this field, the chapter is not intended to cover all aspects of the field but focus on optoelectronic-related application, typically photodiode and phototransistors for photodetection and optoelectronic memories that integrate both light sensing and memory function.

2. Type I, II, and III heterostructures for optoelectronics

The interfacial energy band alignment in heterostructures governs the carrier dynamics in devices and therefore determines directly their functional performances. Depending on the relative positions of conduction band and valance band of constituting materials, there are generally three types of band alignments, including type I (straddling gap), type II (staggered gap), type III (broken gap), as illustrated in **Figure 1a** [21]. The different band offsets make them perform differently in optoelectronic devices [22]. In type I alignment, the bandgap of a semiconductor is located within the bandgap of another one; thus both electrons and holes tend to relax in the first narrow bandgap semiconductor. It is therefore widely used in light emitting diodes for higher light illumination efficiency by confining electron and hole pairs within the narrow bandgap semiconductor [23]. In contrary, in type II alignment, both the conduction band minimum (CBM) and valance band maximum (VBM) are higher or lower than the other, which forces electrons and

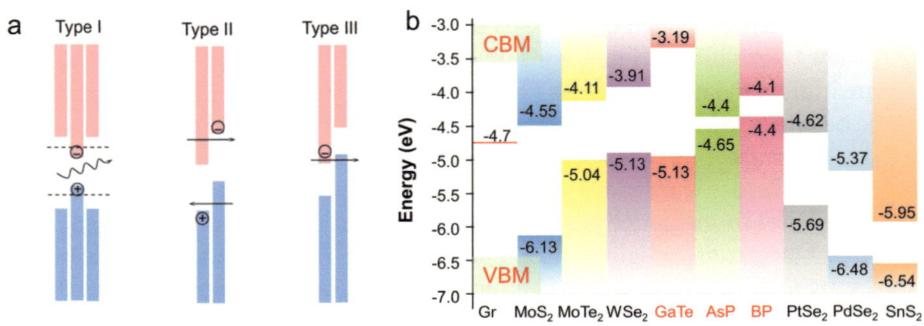

Figure 1.
Band positions and alignments for 2D materials and heterostructures. (a) Heterostructures of type I, II and III interfacial band alignments, reproduced with permission from Ref. [21], Copyright 2016 American Physical Society. (b) Conduction and valance band positions of selected 2D materials collected from literatures.

holes residing in different semiconductors. The separation of electron–hole pairs in type II aligned heterostructures allows the fabrication of rectifying diodes with photovoltaic effects and is usually adopted for photoelectric detectors that transform incident light into electrical signals [3]. In the case of type III band alignment, the bandgap of a semiconductor lies outside of the other one, with its CBM lower than the VBM of the other. There is no more forbidden gap at the interface compared to the bulk semiconductor. Such type III band alignment is useful in tunneling field-effect transistors with large current density [24].

Since the conduction band is usually related to the cations while valance band is related to the anions, designing the band offsets is traditionally mostly achieved in superlattices of semiconductor alloys with widely tuned bandgap and suppressed lattice mismatches, e.g., $Al_xGa_{1-x}As/GaAs$ [25]. However, by using 2D materials, the lattice mismatch between adjacent heterostructured materials is in principle eliminated due to the weak interlayer coupling via van der Waals force. Various 2D materials of different energy band structures and gaps, e.g., graphene, MXenes, black phosphorous (BP), TMDs, and hexagonal boron nitride (h-BN), can thus be artificially stacked to make multiple kinds of heterostructures [13, 26]. **Figure 1b** illustrates the energy bandgap position of several 2D semiconductors [21, 27, 28]. Due to the zero-bandgap characteristic of graphene, it could not be directly used for high on–off switching devices, e.g., transistors, diodes, but is often used as electrode contacts for its ultrahigh carrier mobility >10,000 $cm^2\ V^{-1}\ s^{-1}$ [29]. Recently, other 2D semiconductors have been found as alternatives, with widely distributed bandgaps from 0.2 eV to 2–3 eV [4, 6]. The electron affinity also varies largely from 3 to <5 eV, thus rendering the possibility to make all kinds of heterostructures (types I, II, III) with different band offsets, i.e., by choosing suitable 2D semiconductors. For example, WSe_2/SnS_2 constitutes a type III heterostructure, while MoS_2/WSe_2 forms a typical type II structure. Notably, the number of stacked layers is also not limited to two but can be facilely increased for multilayer heterostructures for tunneling diodes or device encapsulation [13]. The continuously increasing 2D material family thus incubates infinite possibilities in 2D heterostructures and extremely rich functions.

3. Dry transfer methods for artificial 2D stacking

The deterministic transfer of two-dimensional materials constitutes a crucial step toward the fabrication of heterostructures based on the artificial stacking of two-dimensional materials. To stack multiple 2D materials into heterostructure, one needs to transfer 2D materials into a specified position on substrate. This is usually done under an optical microscope, in which one could identify the ultrathin 2D materials through their slight color contrast with substrates. A 3D moving stage is usually equipped to fine adjust the stacking position of each layer, as indicated in **Figure 2a** [30]. So far, a lot of methods and processes have been developed to achieve high-quality assembly of 2D materials in devices and multilayer heterostructures. For 2D materials initially grown on substrates, e.g., graphene on copper, MoS_2 on sapphire, they are usually first etched free from the substrates via polymer (typically poly(methyl methacrylate), known as PMMA)-assisted handling and wet-chemical etching processes [31]. However, the residual of PMMA and wet etching chemicals often deteriorate material performance and also degrades the cleanness of stacked interface, which can be serious in multilayered heterostructures. All-dry transfer of 2D materials is thus desired for making high-performance devices.

To make a heterostructure based on vertical stacking, 2D materials can be typically exfoliated from single crystals by Scotch tape and then transferred to viscous

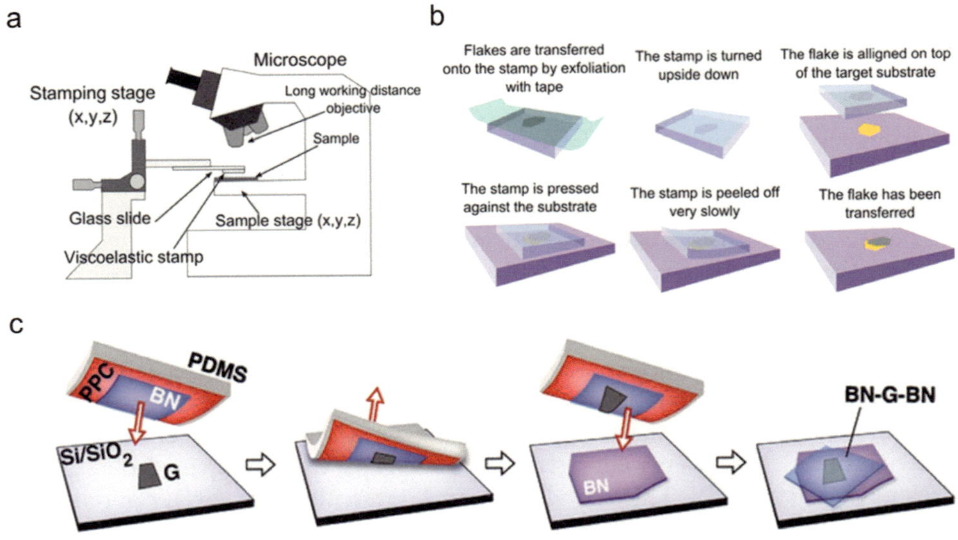

Figure 2.
Setup and typical dry transfer processes for 2D vdW stacking. (a) Schematic of the experimental setup and (b) the processes employed for the all-dry transfer process, reproduced with permission from Ref. [30], Copyright 2014 IOP publishing Ltd. (c) Schematic of the van der Waals (vdW) technique for polymer-free assembly of layered materials, reproduced with permission from Ref. [33], Copyright 2013 Science.

elastomer stamp (poly dimethyl siloxane, PDMS), as illustrated in **Figure 2b** [30]. The transparent PDMS stamp is then used to handle the exfoliated 2D flakes. Under optical microscope, it is then aligned to a target position, e.g., the position of already transferred 2D layer. The position of the stamp is then fine-tuned in all three dimensions to approach the target substrate, until the full contact. It is then slowly peeled off from substrate leaving 2D material behind. Sometimes slight heating of the substrate is necessary to reduce the viscosity of PDMS stamp and promote the successful transfer of 2D material onto target substrate. Instead of the common used PDMS stamps, thermal release tapes can be also used as the handle [32]. Though no wet-chemical etching processes is adopted in the above procedures, the surface of PDMS is full of silane groups and may contaminate the 2D material during transfer and make the contacting substrate hydrophobic. It may therefore deteriorate the material and interface quality in device.

An improved polymer-free method was reported by Wang et al., which adopted the clean h-BN as the buffer layer to attach graphene (**Figure 2c**) [33]. This is based on the stronger interaction between graphene and h-BN compared to SiO_2, so that elastomer stamps with h-BN layer could pick up graphene layer from substrate. Note that the graphene layer is initially transferred onto SiO_2 substrate by tape exfoliation; both the top and bottom surface are free from polymer residuals due to the fresh exfoliation when peeling off the tape. Through this method, the graphene layer during all transfer processes is protected by h-BN and thus could form clean interfaces with both the top and bottom h-BN layers. The as-prepared h-BN-encapsulated graphene manifests unprecedented room temperature mobility up to 140,000 $cm^2\ V^{-1}\ s^{-1}$, with long ballistic transport distance over 15 μm at 40 K, demonstrating the ultrahigh interface quality formed in such polymer-free transfer methods.

Recently, the pick-transfer methods have been also modified to transfer metal electrodes onto 2D materials, avoiding the interdiffusion of elements within the contact interface with 2D materials from traditional physical deposition of metal electrodes (via magnetic sputtering, thermal evaporation, etc.) [34]. Importantly, the formed electrical contact with MoS_2 using different metal electrodes displayed ideal Schottky barriers defined by the work function difference between metal and MoS_2, which have not been achieved in conventional Si devices. It is therefore undoubted that the versatile usages of pick-transfer methods in assembling 2D devices hold vast potential in reforming existing technologies from many aspects.

4. Heterostructures for photodetection

There are several kinds of photodetectors that convert incident light signal to electrical signals, including detectors that rely on photoelectric effect, pyroelectric effect, and photothermal effects. Among the various detectors, the photoelectric detectors exhibit fast response dynamics based on simple separation of electron–hole pairs and are mostly used in commercial products. The photoelectric detectors can be further categorized into photodiode and phototransistors. In photodiodes, the photogenerated electron–hole pairs are separated by the built-in electric field in space charge region, while in phototransistors, an external electric field is applied to generate photodetection gain >100% for highly sensitive detection.

In 2D heterostructures, both photodiodes and phototransistors can be built up by vdW stacking of different materials. Because of the presence of band offset at the interface, heterostructured junctions tend to enable efficient charge separation compared to homojunctions, which requires deliberate control of their p and n doping states. In this section, we discuss several typical heterostructures in type I–III band alignments and their behavior in photodetection.

4.1 Heterostructured diodes

To fabricate heterostructured diode, one kind of 2D material is exfoliated and transferred onto the other one. For the charge separation in vertical direction, type II band alignment is desired. However, this is not naturally obtained, especially when one adopts a narrow bandgap semiconductor for infrared applications, e.g., BP. However, since the work function of ultrathin 2D materials can be dramatically modulated by electrostatic methods, the behavior of 2D diodes was demonstrated tunable by applying gate voltages [35–37]. As illustrated in **Figure 3a**, BP and WSe_2 form type I band alignments [38], with slight conduction band offset (~0.1 eV). When increasing the back-gate voltage from negative to positive, the WSe_2 layer is tuned sequentially from p to i and n states by the injection of gate-coupled electrons and then forming, respectively, p–p, pi, and p-n junctions with the p-typed BP. Further increasing gate bias also tunes BP to n type and results in n-n junction. Accordingly, these junctions manifest different rectification ratios under gate bias. **Figure 3b** displays the forward and reverse channel current (at V_{ds} = 1 V and −1 V); the different onset threshold gate voltage under forward and reverse bias results in a window of −30 V < Vg < 10 V, in which high rectification ratio is obtained by the formation of p-n junction. The widely tuned doping characteristic of 2D bipolar semiconductor thus renders feasible modification of the diode characteristics in the device via various kinds of field effects, including using ionic liquids and ion gels [39].

By choosing appropriate 2D semiconductors, p-n junctions can be formed directly without gate bias. Wang et al. reported such diode based on p-typed gate and n-type MoS_2 [37]. It displays apparent photovoltaic effect under light illumination, as indicated in **Figure 3c**. The extracted ideal factor of the junction is as low as 2 at room temperature, corresponding to Shockley-Read-Hall (SRH) recombination-dominated carrier loss during transport. So far, various kinds of p-n junctions have been made based on such type II band alignments, including BP/MoS_2 [38], MoS_2/$MoTe_2$ [40], MoS_2/WSe_2 [41], etc. The open circuit voltage by photovoltaic effect in such type II

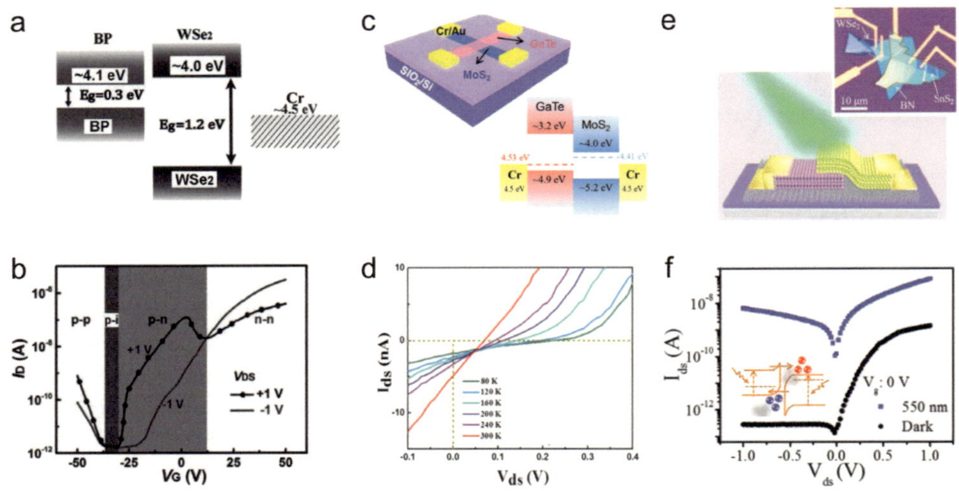

Figure 3.

2D Heterostructures of different interfacial band alignments and their characteristics. (a) The type I band alignment between BP and WSe_2 and (b) the appearance of various junction behaviors (p-p, p-n, n-n) under gate modulation, reproduced with permission from Ref. [38], Copyright 2017 Wiley-VCH. (c) Schematic of the type II heterostructure based on n-type MoS_2 and p-type GaTe, and (d) the photovoltaic characteristic under light irradiation, reproduced with permission from Ref. [37], Copyright 2015 American Chemical Society. (e) Schematic diagram of a type III WSe_2/SnS_2 heterostructure and (f) its IV characteristic under dark and light illumination (550 nm), reproduced with permission from Ref. [19], Copyright 2018 Wiley-VCH.

band heterostructures is limited by the interfacial bandgap determined by the lower conduction and higher valance band. It is therefore usually less than the maximum V_{oc} attainable in p-n junction of each component. However, an essential benefit of such heterostructure is the formation of photodiode without deliberate efforts in controlling the p and n-type doping. A strong evidence of the formation of type II band alignment is that the photoluminescence at the junction area is quenched due to the separation of electron–hole pairs at the interface. Another benefit of such type II heterostructure is based on the interlayer transition, which supports sub-bandgap photodetection [42]. For example, MoS_2/WS_2 heterojunction displays near-infrared response that is beyond both the bandgap limits of MoS_2 and WS_2 [43].

Tunneling diodes can be formed by heterostructures of type III band alignment [24]. In the case of WSe_2/SnS_2 heterostructure, due to the high electron affinity of SnS_2, type III heterostructure is formed with direct interband transition between valance band of WSe_2 and the conduction band of SnS_2 [19]. The diode initially displayed high rectification ratio $>10^4$ for low dark current under reverse bias, whereas under illumination, the device exhibits dramatically increased light current by direct tunneling, resulting in high responsivity >200 AW^{-1} and excellent detectivity $>10^{13}$ Jones. Further exploration of the kind of heterostructure using other 2D materials with different bandgap may have the potential to make high-performance tunneling photodiodes for infrared. The heterostructure of narrow bandgap BP and larger bandgap MoS_2 has been used to realize multi-value inverters with high gains >150 based on gate-modulated tunneling current [44].

In addition to the two-layer stacking, multilayered heterostructures have been also developed as tunneling diodes. **Figure 4a** illustrates such a heterostructure based on vertically stacked graphene/MoS_2/graphene [35]. Because of the work function between top and bottom graphene (due to the unidentical substrate doping effect), the multilayer displayed photovoltaic separation of electron–hole pairs under illumination, reaching a V_{oc}~0.3 V under additional gate bias. The device also exhibits wavelength-dependent responsivity that is related to the absorption in

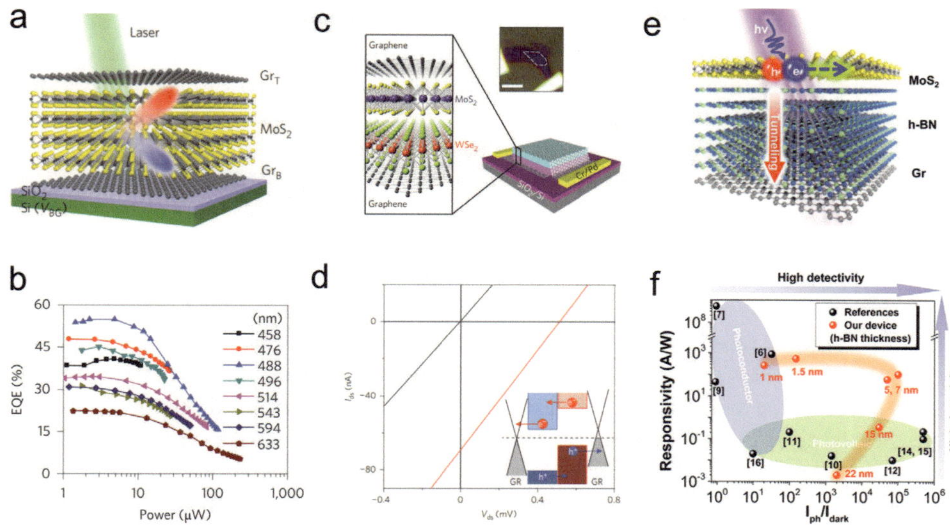

Figure 4.
Various kinds of vertical heterostructures. (a) Schematic of Gr/MoS_2/Gr heterojunction, which displays photovoltaic separation of electron–hole pairs and (b) the extracted external quantum efficiency (EQE) under different light power and wavelengths, reproduced with permission from Ref. [35], Copyright 2013 Nature Publishing Group. (c) The schematic diagram of vertical p-n junction made by MoS_2/WSe_2 sandwiched within two graphene layers and (d) their IV characteristic in dark and illumination, reproduced with permission from Ref. [36], Copyright 2014 Nature Publishing Group. (e) Schematic of a tunneling diode based on graphene/h-BN/MoS_2 heterostructure and (f) its photodetection performance when using h-BN with thickness within 1–22 nm, reproduced with permission from Ref. [18], Copyright 2016 American Chemical Society.

MoS_2 (as indicated in **Figure 4b**), demonstrating the working principle of the multilayer junction. By using graphene at both the bottom and top of the junction made by MoS_2 and WSe_2, Lee et al. demonstrated an efficient p-n junction at the ultimate atomic thin thickness with improved collection of photoexcited carriers [36]. **Figure 4c** and **d** illustrate the structure and *IV* characteristic of the device. Under illumination, tunneling-assisted interlayer recombination of the majority carriers dominates the electronic and optoelectronic behavior of the junction. Alternatively, sandwiching graphene within WSe_2 and MoS_2 can make a broadband photodetector up to 2 μm based on the absorption of graphene [45]. Such interlayer tunneling can be suppressed by inserting an insulating h-BN layer. Vu et al. fabricated a tunneling heterostructure based on graphene/h-BN/MoS_2 in the configuration shown in **Figure 4e** [18]. The dark current in device is greatly suppressed by blocking direct tunneling. However, under illumination, photogenerated carriers may overcome the barrier and contribute significant photocurrent via Fowler-Nordheim (FN) tunneling. Notably, to balance the photodetection performance in terms of the responsivity and detectivity, the thickness of h-BN is optimal ~5–7 nm, as indicated in **Figure 4f**. The thicker the h-BN layer, the lower probability of FN tunneling and thus lesser photocurrent, while the thinner h-BN results in large dark current by direct tunneling, therefore less detectivity in photodetection.

4.2 Heterostructured phototransistors

In photodiodes, the photodetection gain is limited due to the maximum attainable quantum efficiency (photon-to-electron conversion efficiency) less than unity [46]. Hence, photodiodes are less sensitive and are usually operated under reverse bias or self-driven mode without external bias. In comparison, when integrating such heterostructure into a photoconductor configuration, phototransistors can be made with high sensitivity based on the photoconductive gain and vertical photovoltaic effects. The photodetection gain originates from the separation of electron–hole pairs at the heterostructure interface, with one kind of carrier accumulated in the 2D high mobility channel, therefore yielding amplified photoconductive gains by the ratio of injected charges compared to the inherent carrier concentration in 2D channel [47]. A representative example is PbS quantum dot (QD)-sensitized graphene, in which the QDs and 2D surface are coupled by vdW interaction (**Figure 5a**) [48]. Upon illumination, holes are injected into graphene and transport there with dramatically increased mobility compared to QDs that have large amount of grain boundaries and surface states. In this way, ultrahigh responsivity >10^7 A/W has been demonstrated in such hybrid photodetectors. Notably, based on the gate-modulated Fermi level in graphene, the charge injection from PbS QDs to graphene can be extensively tailored. As indicated in **Figure 5b**, the attained responsivity is sensitive to the applied gate bias; under V_g = 4 V, the photoresponse gain is tuned even to zero by eliminating the interfacial charge transfer. Such widely tuned gain is potentially useful for intentionally selected sensitivity levels for a detector. However, due to the zero-bandgap nature of graphene, hybrid detectors with graphene as the channel exhibit large dark current and low detectivity. Alternatively, other 2D semiconductors, such as MoS_2 and WSe_2, have been also explored as the channel, yielding improved on–off ratio in detector [47, 49].

In addition to colloidal quantum dots, 2D heterostructures based on vertically stacked 2D layers can also make up phototransistors. A narrow bandgap semiconductor can be stacked on another 2D material for extended photodetection spectra. As illustrated in **Figure 5c**, BP is stacked on a WSe_2 channel [17]. The photoexcited carriers in BP by near-infrared photons are separated by the type II interface, with

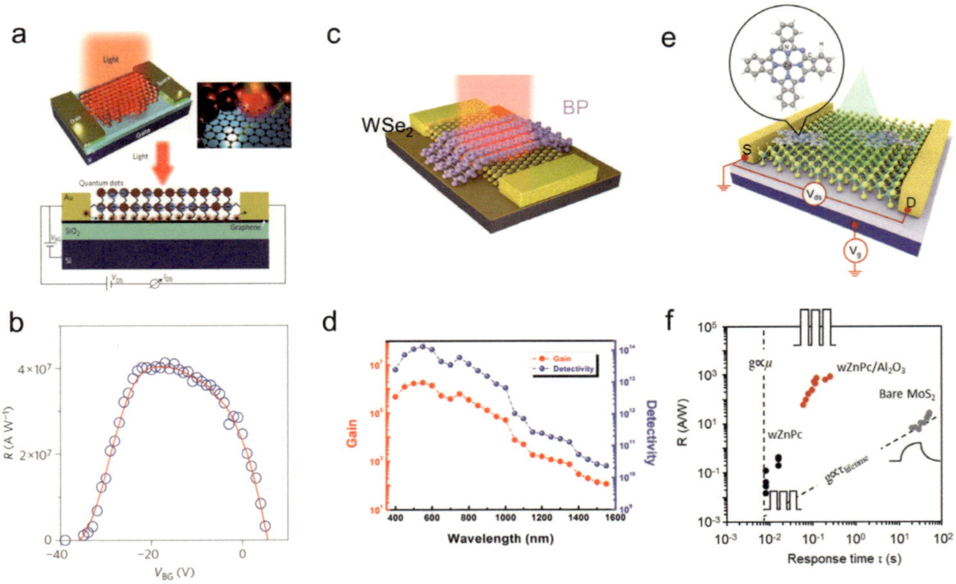

Figure 5.
Phototransistors based on various heterostructures. (a) The schematic of PbS quantum dots sensitized graphene for infrared photodetection; (b) the back-gate-modulated responsivity of the hybrid photodetector, reproduced with permission from Ref. [48], Copyright 2012 Nature Publishing Group. dependence of the responsivity on the different wavelength. (c) Configuration of a vertically stacked BP/WSe_2 heterostructure and (d) its wavelength-dependent gain and detectivity, reproduced with permission from Ref. [17], Copyright 2017 Elsevier Ltd. (d) Detectivity of various photodetector versus wavelength of the incident laser. (e) Illustration of the organic/inorganic vdW heterostructured phototransistor based on ZnPc-decorated MoS_2 and (f) its photoresponse behavior, which is greatly improved compared to photoconductors that suffer persistent photoconductance, reproduced with permission from Ref. [16], Copyright 2018 American Chemical Society.

electrons injected to WSe_2. The amount of injected charge is related to the junction capacitance and the photovoltage built across the junction. In **Figure 5d**, the photodetection gain in such device reaches 10^2 at 1500 nm, which is considerably larger than the photodiodes (<1) by the amplification mechanism in phototransistor. Therefore, the specific detectivity of the device reaches 10^{10}–101^4 Jones at the measured wavelength (400–1500 nm) range. Longer wavelength results in low gain and detectivity due to the decrease of light absorption. Instead of BP, a lot of other 2D materials has been also explored to construct such heterostructured phototransistor, in which the photovoltaic separation of photocarriers can be used to gate the semiconductor channel and amplify the photoconductive gain.

Without complicated stacking processes, 2D vdW heterostructures can be also made by combining organic small molecules with 2D material. As illustrated in **Figure 5e**, Huang et al. recently reported such a vdW phototransistor based on Zinc phthalocyanine (ZnPc, a π-conjugated planar molecule)-decorated monolayer MoS_2, which is achieved by simple solution treatment [16]. The formed junction displayed apparent rectification characteristic at the out-plane direction by forming type II band alignment and p-n junction. As a result, the detector displayed remarkably improved response speed and optimal responsivity (**Figure 5f**) with proper Al_2O_3 passivation. Other molecules, such as pentacene, have been also used to modify the performance of 2D semiconductors (MoS_2, ReS_2, etc.) in addition to response dynamics but also the response spectra [50, 51]. Considering the huge library of 2D materials and organic molecules, it is believed such hybrid heterostructure holds special promise in achieving scalable high-performance photodetections, in which using existing pick-transfer procedures is apparently challenging.

5. Optoelectronic memories

Optoelectronic memory can transform incident optical signals into stored electric charges [52]. Considering the light program signals can be free from interferences, the optoelectronic memories are particularly attractive for realizing high-throughput data storage, e.g., in parallel computing [53]. A typical optoelectronic memory is consisted of light sensing part and charge storage component, which could be feasibly realized using multilayered 2D stacking. Compared to the conventional 3D counterparts, the 2D devices have the advantages in having high on–off ratio by the ultrathin channel, the conductance of which can be feasibly modulated via slight amount of trapped charges. According to the charge trapping mechanism, in the following we describe two kinds of optoelectronic memories, based on, respectively, the charge trapping in (i) defect energy states or (ii) float gates.

5.1 Charge trapping in defect levels

The ultrathin nature makes 2D semiconductors highly suitable as the readout channel in memory, as their conductance can be modulated greatly by slight charge trapping, including by the inherent trap states in devices. In literatures, the prepared MoS_2 often exhibits midgap trap states [54], and the device also suffers from interface defect states, e.g., at the interface with SiO_2 [55], which may capture some charges under gate modulation by the shifted Fermi level E_F (the trap states below E_F are prone to be filled with electrons, while those states above E_F tend to be empty). This usually results in large hysteresis in field-effect devices and different conduction states after positive and negative gate stress. However, the limited density of trap states restricted the on–off switch in memory. Lee et al. reported an improved device by introducing localized electronic states in MoS_2 using tailored

SiO_2 substrate with functional silanol groups (Si-OH) (**Figure 6a**) [56], which exhibit strong polar interaction and causes local potential fluctuation in energy band. The device is composed of thin MoS_2 layer on SiO_2 substrate, using the back-Si as the gate. The conduction state is reset by using positive gate bias (80 V) and then programmed using light exposure under gate bias (20 V). Applying V_G = 80 V fills the traps with electrons, resulting in *OFF* states of channel when removing V_G, while light exposure releases the trapped electrons by generating electron–hole pairs that promote the charge release. The device manifested highly linear readout charges programmed by light exposure time. However, since the trapped charges can be thermally activated to conduction/valanced band for trapped electrons/holes, the programmed states exhibit transient change of conduction states after initial program (**Figure 6b**), and the charge readout is slow ~seconds.

The above optoelectronic memory works under visible light excitation due to the bandgap limit of MoS_2. Wang et al. reported an infrared memory using the vdW heterostructure of MoS_2/PbS [57], which is sensitive to 1550 nm radiation with the sensitization of narrow bandgap PbS thin flakes epitaxially grown on MoS_2 (as illustrated in **Figure 6c**). The charge trapping is based on the electron injection into MoS_2 by the generation of large amount of photoexcited electrons in PbS under light illumination, as indicated by the energy band diagram shown in **Figure 6d**. However, the device exhibits low resistance change by light exposure and transient conductance variation after program, due to the eventual recombination of electron–hole pairs in dark, which drive carrier distribution to equilibrium. Also, the program speed is directly determined by photon energy and the overall incident power, as the former governs the energy of photoexcited carriers (whether it is sufficient to overcome the interfacial potential barrier to be injected into the other side) and the latter determines the number of excited carriers. Alternatively, the charge trapping in defect states in dielectric materials tends to exhibit long retention time. Xiang et al. constructed a nonvolatile memory using WSe_2 transferred on insulate h-BN layer (**Figure 6e**) [58]. The inherent defect states in h-BN are able

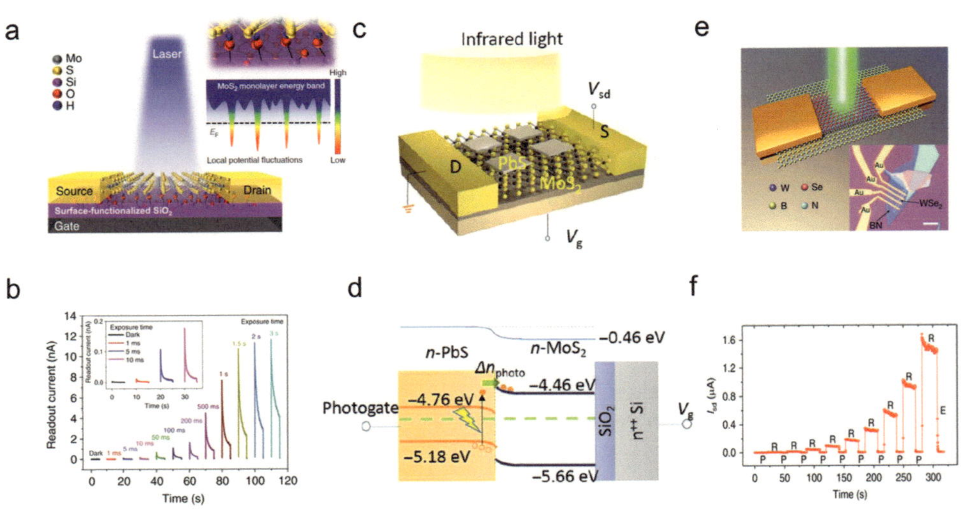

Figure 6.
2D optoelectronic memories. (a) Schematic of a 2D memory based on MoS_2 on tailored SiO_2 surface and (b) the light exposure time programmed memory states in the device, reproduced with permission from Ref. [56], Copyright 2017 Nature Publishing Group. (c) An infrared memory based on vdW heterostructure of PbS/MoS_2 and (d) the schematic of its band alignment, reproduced with permission from Ref. [57], Copyright 2018 Science. (e) The schematic and optical image (inset) of an optoelectronic memory based on WSe_2 on h-BN, (f) shows the device conductance change during electrical erase and optical program, reproduced with permission from Ref. [58], Copyright 2018 Nature Publishing Group.

to trap photoexcited carriers in WSe_2, therefore enabling optoelectronic memory operation. The memory is operated under the simultaneous light exposure and gate bias, thus to force the charge trapping into the midgap states of h-BN. Because of the large bandgap of h-BN, the trapped charges can hardly move, and the resulted memory exhibits long-term retention characteristics for more than 10^4 s. Such optoelectronic memory can be feasibly transformed into multi-bit memory, by using either the amplitude of gate bias or the light irradiation power, wavelength, and pulse number as the input (**Figure 6f**). However, slight temporal change of conductance is still observed due to the recombination of photogenerated electron–hole pairs in WSe_2 itself. Nevertheless, the strategy has been exploited to develop artificial optoelectronic synapses, the overall weight of which is less sensitive to the single-unit device but to the average of multiple connections [59].

5.2 Float gate heterostructures

Instead of charge trapping in random trap states, float gate structure exhibits well-described charge trapping characteristics and long-term retention characteristics [60]. The charge trapping can also be triggered by light irradiation to the light sensing semiconductor channel or float gate. Using 2D materials, the float gate structure can be assembled by h-BN as the insulate barrier and 2D semiconductors as the channel.

Figure 7a displays the initial 2D float memory based on graphene and MoS_2 separated by h-BN [61]. The device usually has the structure of a field-effect transistor but with an additional float gate inserted between the source-drain channel and the control gate. The memory behavior of the device by using MoS_2 as the channel and graphene as the float gate is shown in **Figure 7b**. The charge trapping is based on the quantum tunneling under gate bias, which induces FN tunneling by lowering the effective tunneling barrier with trigonal potential profile in the insulate h-BN

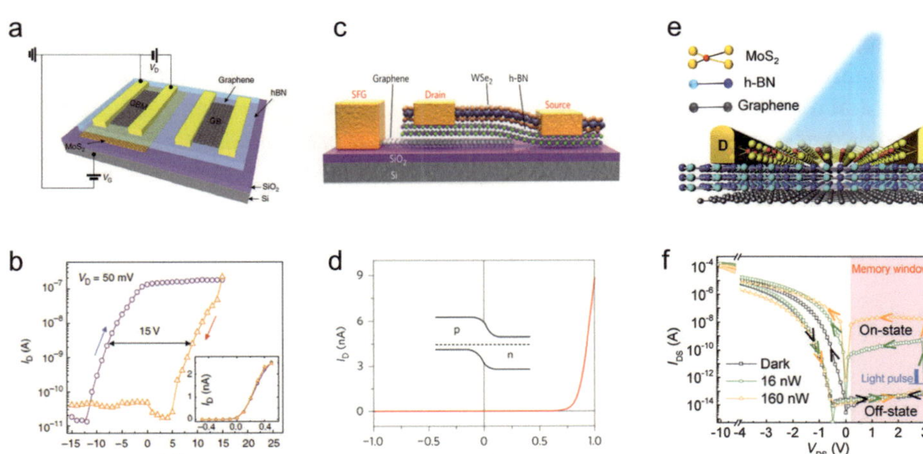

Figure 7.
Several representative 2D float gate heterostructures. (a) The schematic configuration of a float gate memory based on MoS_2/h-BN/graphene (b) shows the hysteresis memory behavior using back-Si gate, and the inset depicts the MoS_2 conductance modulated by float gate potential, reproduced with permission from Ref. [61], Copyright 2013 Nature Publishing Group. (c) Schematic of the semi-float gate device base on graphene/h-BN/WSe_2 heterostructure for the formation of lateral diode and (d) its IV characteristic showing rectification behavior by the formation of p-n junction, reproduced with permission from Ref. [62], Copyright 2017 Nature Publishing Group. (e) A two-terminal optoelectronic memory based on vdW heterostructure of MoS_2/h-BN/graphene (f) displays its IV characteristics in dark and light illumination, showing the light programmed on and off states under positive bias, and the memory states are electrically erased using large negative bias, reproduced with permission from Ref. [20], Copyright 2017 Wiley-VCH.

layer. Alternatively using graphene as the channel results in low on–off ratio due to the zero-bandgap nature of graphene, by what the graphene channel can hardly be turned off. Notably, the thickness of h-BN is critical for the float memory, as too thin h-BN results in direct tunneling loss of charges and poor retention behavior, while too thick h-BN is good for retention but requires high operation voltages. The optimal thickness of h-BN is ~6–10 nm. The thin thickness of h-BN enables efficient tunneling of channel conductance by the float gate potential, as indicated in the inset of **Figure 7b**. Instead of graphene and MoS_2, many other 2D semiconductors have been explored for the float memory, including WSe_2, ReS_2, BP, etc. [63–66]. They all displayed high on–off ratio up to 10^7, which is likely to benefit multi-bit storage.

Because of the excellent tunability of charges in 2D channel, the float gate structure has been reformed into semi-float and two-terminal structures. **Figure 7c** shows a semi-float gate device with WSe_2 as the channel [62], in which the graphene as float gate spans half of the channel. Thus, the charge trapping in graphene only modulates the carrier concentration in partially the overlapped region. Taking advantage of the ambipolar characteristic of WSe_2, the gate region can be tuned either p- or n-doped, forming the lateral pn diodes or Schottky diodes with apparent rectification behavior (**Figure 7d**). A special advantage of such device is their reconfigurable device behavior on demand. The device structure can be further simplified into two-terminal structures by removing the control gate, which usually is the back-Si gate [20]. **Figure 7e** displays a schematic structure of such two-terminal float memory. The charge tunneling can be realized by applying enough source-drain bias as indicated in **Figure 7f**. Because of the nonuniform electric field in channel, the potential drop between drain and float gate is sufficient to induce charge injection into float gate. After applying negative V_{ds}, electrons are injected into graphene, resulting in off state when reading at $V_{ds} > 0$. However, shining light to the device releases the trapped charges and recovers the initial state. Thus, the memory can be electrically erased and programmed by light exposure. Due to the absorption limit of MoS_2, the device is only programmable with wavelengths <650 nm. By controlling the light dose with power and duration, the device manifests 18 states, rendering potential application for multi-bit purposes. However, an essential drawback of such two-terminal device is the high power consumption during electrical erase, as high source-drain current is present compared to the negligible leakage current via gate coupling.

The various heterostructures by versatile 2D stacking have enabled the blossom of 2D optoelectronic devices. There is also an emerging of optoelectronic programmed logic elements using the flexible gate coupling in ultrathin thickness [67]. The pathway toward multifunctional 2D devices seems very promising to stimulate indispensable applications based on continuously expanding family of 2D materials.

6. Conclusions

As a summary, in this chapter, we have introduced various types of 2D heterostructures for both photodetection and optoelectronic memory, both of which extensively take advantage of the feasible field-effect modulation to the optoelectronic properties of 2D materials. In the past few years, we have witnessed the marvelous revolution of design and construction of functional devices using diverse 2D materials and feasible vdW stacking methods. The progress will undoubtedly continue given the remarkable flexibility of stacking 2D material in atomic thickness, which had been extremely challenging for 3D materials. However, one shall expect critical breakthroughs are necessary before their practical applications, especially in the large-scale fabrication of vdW devices, and the development of indispensable functions compared to the existing ones in consumer markets.

Acknowledgements

This work was supported by National Natural Science Foundation of China (Grant No. 61804059).

Conflict of interest

The authors declare no conflict of interest.

Author details

Hongcheng Ruan[1], Yu Huang[2], Yuqian Chen[2] and Fuwei Zhuge[2*]

1 School of Information and Safety Engineering, Zhongnan University of Economics and Law, Wuhan, China

2 State Key laboratory of Materials Processing and Die and Mould Technology, School of Materials Science and Engineering, Huazhong University of Science and Technology (HUST), Wuhan, China

*Address all correspondence to: zhugefw@hust.edu.cn

References

[1] Novoselov KS, Geim AK, Morozov SV, Jiang D, Zhang Y, Dubonos SV, et al. Electric field effect in atomically thin carbon films. Science. 2004;**306**:666-669

[2] Radisavljevic B, Radenovic A, Brivio J, Giacometti V, Kis A. Single-layer MoS_2 transistors. Nature Nanotechnology. 2011;**6**:147-150

[3] Koppens FHL, Mueller T, Avouris P, Ferrari AC, Vitiello MS, Polini M. Photodetectors based on graphene, other two-dimensional materials and hybrid systems. Nature Nanotechnology. 2014;**9**:780-793

[4] Butler SZ, Hollen SM, Cao L, Cui Y, Gupta JA, Gutierrez HR, et al. Progress, challenges, and opportunities in two-dimensional materials beyond graphene. ACS Nano. 2013;**7**:2898-2926

[5] Fiori G, Bonaccorso F, Iannaccone G, Palacios T, Neumaier D, Seabaugh A, et al. Electronics based on two-dimensional materials. Nature Nanotechnology. 2014;**9**:768-779

[6] Wang QH, Kalantar-Zadeh K, Kis A, Coleman JN, Strano MS. Electronics and optoelectronics of two-dimensional transition metal dichalcogenides. Nature Nanotechnology. 2012;**7**:699-712

[7] Chhowalla M, Shin HS, Eda G, Li LJ, Loh KP, Zhang H. The chemistry of two-dimensional layered transition metal dichalcogenide nanosheets. Nature Chemistry. 2013;**5**:263-275

[8] Zhang YJ, Oka T, Suzuki R, Ye JT, Iwasa Y. Electrically switchable chiral light-emitting transistor. Science. 2014;**344**:725-728

[9] Sundaram RS, Engel M, Lombardo A, Krupke R, Ferrari AC, Avouris P, et al. Electroluminescence in single layer MoS_2. Nano Letters. 2013;**13**:1416-1421

[10] Huang C, Du Y, Wu H, Xiang H, Deng K, Kan E. Prediction of intrinsic ferromagnetic ferroelectricity in a transition-metal halide monolayer. Physical Review Letters. 2018;**120**:147601

[11] Cui C, Hu WJ, Yan X, Addiego C, Gao W, Wang Y, et al. Intercorrelated in-plane and out-of-plane ferroelectricity in ultrathin two-dimensional layered semiconductor In_2Se_3. Nano Letters. 2018;**18**:1253-1258

[12] Luo P, Zhuge FW, Zhang QF, Chen YQ, Lv L, Huang Y, et al. Doping engineering and functionalization of two-dimensional metal chalcogenides. Nanoscale Horizons. 2019;**4**:26-51

[13] Geim AK, Grigorieva IV. Van der Waals heterostructures. Nature. 2013;**499**:419-425

[14] Cui X, Lee GH, Kim YD, Arefe G, Huang PY, Lee CH, et al. Multi-terminal transport measurements of MoS_2 using a van der Waals heterostructure device platform. Nature Nanotechnology. 2015;**10**:534-540

[15] Ma Q, Andersen TI, Nair NL, Gabor NM, Massicotte M, Lui CH, et al. Tuning ultrafast electron thermalization pathways in a van der Waals heterostructure. Nature Physics. 2016;**12**:455-459

[16] Huang Y, Zhuge F, Hou J, Lv L, Luo P, Zhou N, et al. Van der Waals coupled organic molecules with monolayer MoS_2 for fast response photodetectors with gate-tunable responsivity. ACS Nano. 2018;**12**:4062-4073

[17] Ye L, Wang P, Luo W, Gong F, Liao L, Liu T, et al. Highly polarization

sensitive infrared photodetector based on black phosphorus-on-WSe_2 photogate vertical heterostructure. Nano Energy. 2017;**37**:53-60

[18] Vu QA, Lee JH, Nguyen VL, Shin YS, Lim SC, Lee K, et al. Tuning carrier tunneling in van der Waals heterostructures for ultrahigh detectivity. Nano Letters. 2017;**17**:453-459

[19] Zhou X, Hu X, Zhou S, Song H, Zhang Q, Pi L, et al. Tunneling diode based on WSe_2/SnS_2 heterostructure incorporating high detectivity and responsivity. Advanced Materials. 2018;**30**:1703286

[20] Tran MD, Kim H, Kim JS, Doan MH, Chau TK, Vu QA, et al. Two-terminal multibit optical memory via van der Waals heterostructure. Advanced Materials. 2019;**31**:e1807075

[21] Özçelik VO, Azadani JG, Yang C, Koester SJ, Low T. Band alignment of two-dimensional semiconductors for designing heterostructures with momentum space matching. Physical Review B. 2016;**94**:035125

[22] Neamen DA. Semiconductor Physics and Devices: Basic Principles. Homewood, IL: Irwin; 1992

[23] Ponce FA, Bour DP. NItride-based semiconductors for blue and green light-emitting devices. Nature. 1997;**386**:351-359

[24] Yan R, Fathipour S, Han Y, Song B, Xiao S, Li M, et al. Esaki diodes in van der Waals heterojunctions with broken-gap energy band alignment. Nano Letters. 2015;**15**:5791-5798

[25] Vurgaftman I, Meyer JR, Ram-Mohan LR. Band parameters for III-V compound semiconductors and their alloys. Journal of Applied Physics. 2001;**89**:5815-5875

[26] Novoselov KS, Mishchenko A, Carvalho A, Castro Neto AH. 2D materials and van der Waals heterostructures. Science. 2016;**353**:aac9439

[27] Rasmussen FA, Thygesen KS. Computational 2D materials database: Electronic structure of transition-metal dichalcogenides and oxides. Journal of Physical Chemistry C. 2015;**119**:13169-13183

[28] Zhuang HL, Hennig RG. Computational search for single-layer transition-metal dichalcogenide photocatalysts. Journal of Physical Chemistry C. 2013;**117**:20440-20445

[29] Yamoah MA, Yang W, Pop E, Goldhaber-Gordon D. High-velocity saturation in graphene encapsulated by hexagonal boron nitride. ACS Nano. 2017;**11**:9914-9919

[30] Castellanos-Gomez A, Buscema M, Molenaar R, Singh V, Janssen L, van der Zant HSJ, et al. Deterministic transfer of two-dimensional materials by all-dry viscoelastic stamping. 2D Materials. 2014;**1**:011002

[31] Chen Y, Gong XL, Gai JG. Progress and challenges in transfer of large-area graphene films. Advanced Science. 2016;**3**:1500343

[32] Pizzocchero F, Gammelgaard L, Jessen BS, Caridad JM, Wang L, Hone J, et al. The hot pick-up technique for batch assembly of van der Waals heterostructures. Nature Communications. 2016;**7**:11894

[33] Wang L, Meric I, Huang PY, Gao Q, Gao Y, Tran H, et al. One-dimensional electrical contact to a two-dimensional material. Science. 2013;**342**:614-617

[34] Liu Y, Guo J, Zhu E, Liao L, Lee SJ, Ding M, et al. Approaching the Schottky-Mott limit in van der Waals

metal-semiconductor junctions. Nature. 2018;**557**:696-700

[35] Yu WJ, Liu Y, Zhou H, Yin A, Li Z, Huang Y, et al. Highly efficient gate-tunable photocurrent generation in vertical heterostructures of layered materials. Nature Nanotechnology. 2013;**8**:952-958

[36] Lee CH, Lee GH, van der Zande AM, Chen W, Li Y, Han M, et al. Atomically thin p-n junctions with van der Waals heterointerfaces. Nature Nanotechnology. 2014;**9**:676-681

[37] Wang F, Wang Z, Xu K, Wang F, Wang Q, Huang Y, et al. Tunable GaTe-MoS_2 van der Waals p-n Junctions with novel optoelectronic performance. Nano Letters. 2015;**15**:7558-7566

[38] Li D, Wang B, Chen M, Zhou J, Zhang Z. Gate-controlled BP-WSe_2 heterojunction diode for logic rectifiers and logic optoelectronics. Small. 2017;**13**:1603726

[39] Choi Y, Kang J, Jariwala D, Kang MS, Marks TJ, Hersam MC, et al. Low-voltage complementary electronics from ion-gel-gated vertical van der Waals heterostructures. Advanced Materials. 2016;**28**:3742-3478

[40] Duong NT, Lee J, Bang S, Park C, Lim SC, Jeong MS. Modulating the functions of MoS_2/$MoTe_2$ van der Waals heterostructure via thickness variation. ACS Nano. 2019;**13**:4478-4485

[41] Doan M-H, Jin Y, Adhikari S, Lee S, Zhao J, Lim SC, et al. Charge transport in MoS_2/WSe_2 van der Waals heterostructure with tunable inversion Layer. ACS Nano. 2017;**11**:3832-3840

[42] Chen H, Wen X, Zhang J, Wu T, Gong Y, Zhang X, et al. Ultrafast formation of interlayer hot excitons in atomically thin MoS_2/WS_2 heterostructures. Nature Communications. 2016;**7**:12512

[43] Wang G, Li L, Fan W, Wang R, Zhou S, Lü J-T, et al. Interlayer coupling induced infrared response in WS_2/MoS_2 heterostructures enhanced by surface plasmon resonance. Advanced Functional Materials. 2018;**28**:1800339

[44] Huang M, Li S, Zhang Z, Xiong X, Li X, Wu Y. Multifunctional high-performance van der Waals heterostructures. Nature Nanotechnology. 2017;**12**:1148-1154

[45] Long M, Liu E, Wang P, Gao A, Xia H, Luo W, et al. Broadband photovoltaic detectors based on an atomically thin heterostructure. Nano Letters. 2016;**16**:2254-2259

[46] Zhuge F, Zheng Z, Luo P, Lv L, Huang Y, Li H, et al. Nanostructured materials and architectures for advanced infrared photodetection. Advanced Materials Technologies. 2017;**2**:1700005

[47] Kufer D, Konstantatos G. Photo-FETs: Phototransistors enabled by 2D and 0D nanomaterials. ACS Photonics. 2016;**3**:2197-2210

[48] Konstantatos G, Badioli M, Gaudreau L, Osmond J, Bernechea M, Garcia de Arquer FP, et al. Hybrid graphene-quantum dot phototransistors with ultrahigh gain. Nature Nanotechnology. 2012;**7**:363-368

[49] Kufer D, Nikitskiy I, Lasanta T, Navickaite G, Koppens FHL, Konstantatos G. Hybrid 2D-0D MoS_2-PbS quantum dot photodetectors. Advanced Materials. 2015;**27**:176-180

[50] Jariwala D, Howell SL, Chen KS, Kang J, Sangwan VK, Filippone SA, et al. Hybrid, gate-tunable, van der Waals p–n heterojunctions from pentacene and MoS_2. Nano Letters. 2016;**16**:497-503

[51] Jiang J, Ling C, Xu T, Wang W, Niu X, Zafar A, et al. Defect engineering for modulating the trap states in 2D

photoconductors. Advanced Materials. 2018;**30**:e1804332

[52] Zhai YB, Yang JQ, Zhou Y, Mao JY, Ren Y, Roy VAL, et al. Toward non-volatile photonic memory: Concept, material and design. Materials Horizons. 2018;**5**:641-654

[53] Gu M, Li X, Cao Y. Optical storage arrays: A perspective for future big data storage. Light: Science and Applications. 2014;**3**:e177

[54] Late DJ, Liu B, Matte HSSR, Dravid VP, Rao CNR. Hysteresis in single-layer MoS_2 field effect transistors. ACS Nano. 2012;**6**:5635-5641

[55] Kaushik N, Mackenzie DMA, Thakar K, Goyal N, Mukherjee B, Boggild P, et al. Reversible hysteresis inversion in MoS_2 field effect transistors. npj 2D Materials and Applications. 2017;**1**:34

[56] Lee J, Pak S, Lee YW, Cho Y, Hong J, Giraud P, et al. Monolayer optical memory cells based on artificial trap-mediated charge storage and release. Nature Communications. 2017;**8**:14734

[57] Wang Q, Wen Y, Cai K, Cheng R, Yin L, Zhang Y, et al. Nonvolatile infrared memory in MoS_2/PbS van der Waals heterostructures. Science Advances. 2018;**4**:eaap7916

[58] Xiang D, Liu T, Xu J, Tan JY, Hu Z, Lei B, et al. Two-dimensional multibit optoelectronic memory with broadband spectrum distinction. Nature Communications. 2018;**9**:2966

[59] Seo S, Jo SH, Kim S, Shim J, Oh S, Kim JH, et al. Artificial optic-neural synapse for colored and color-mixed pattern recognition. Nature Communications. 2018;**9**:5106

[60] Wang SP, He CL, Tang J, Lu XB, Shen C, Yu H, et al. New floating gate memory with excellent retention characteristics. Advanced Electronic Materials. 2019;**5**:1800726

[61] Choi MS, Lee GH, Yu YJ, Lee DY, Lee SH, Kim P, et al. Controlled charge trapping by molybdenum disulphide and graphene in ultrathin heterostructured memory devices. Nature Communications. 2013;**4**:1624

[62] Liu C, Yan X, Song X, Ding S, Zhang DW, Zhou P. A semi-floating gate memory based on van der Waals heterostructures for quasi-non-volatile applications. Nature Nanotechnology. 2018;**13**:404-410

[63] Wang Y, Liu E, Gao A, Cao T, Long M, Pan C, et al. Negative photoconductance in van der Waals heterostructure-based floating gate phototransistor. ACS Nano. 2018;**12**:9513-9520

[64] Gong F, Luo W, Wang J, Wang P, Fang H, Zheng D, et al. High-sensitivity floating-gate phototransistors based on WS_2 and MoS_2. Advanced Functional Materials. 2016;**26**:6084-6090

[65] Feng Q, Yan F, Luo W, Wang K. Charge trap memory based on few-layer black phosphorus. Nanoscale. 2016;**8**:2686-2692

[66] Li D, Wang X, Zhang Q, Zou L, Xu X, Zhang Z. Nonvolatile floating-gate memories based on stacked black phosphorus-boron nitride-MoS_2 heterostructures. Advanced Functional Materials. 2015;**25**:7360-7365

[67] Liu C, Chen H, Hou X, Zhang H, Han J, Jiang YG, et al. Small footprint transistor architecture for photoswitching logic and in situ memory. Nature Nanotechnology. 2019;**14**:662-667. DOI: 10.1038/s41565-019-0462-6

Chapter 10

Catalytic Micro/Nanomotors: Propulsion Mechanisms, Fabrication, Control, and Applications

Liangxing Hu, Nan Wang and Kai Tao

Abstract

Micro-/nanomotors are self-propelled micro-/nanomachines, which are capable of converting the surrounding fuels into mechanical movement or force. Inspired by naturally occurring biomolecular motor proteins, scientists extensively paid great attentions to synthetic micro-/nanomotors. Especially, a number of researchers devoted their efforts onto catalytic micro-/nanomotors. In the past few decades, several advanced developments and excellent contributions have been made in catalytic micro-/nanomotors. The future of this research field can be bright, but some major existing challenges such as biocompatible materials and fuels, smart controlling, and specifically practical applications are still required to be resolved. Therefore, it is essential for us to learn the state of the art of catalytic micro-/nanomotors. In this chapter, the propulsion mechanisms, fabrication methods, controlling strategies, and potential applications of catalytic micro-/nanomotors are presented and summarized.

Keywords: catalytic micro-/nanomotors, propulsion mechanisms, fabrication methods, controlling strategies, applications

1. Introduction

Movement is crucial for all different types of lives existing in both macroscopic and micro-/nanoscopic environments. Nature has developed smart and high-efficiency biomolecular motor proteins through such many years of biological evolution and has employed them in numerous biological processes and cellular activities [1–3]. For instance, bacteria are able to propel themselves forward with the aid of rotary flagellar nanomotors, as shown in **Figure 1A** [1]. Moreover, linear biomolecular motor proteins, such as kinesin, myosin, and dynein, are capable of harvesting energy from hydrolyzing adenosine triphosphate (ATP) into adenosine diphosphate (ADP) and phosphate (Pi) molecule for lateral motion along corresponding tracts, as shown in **Figure 1B** [2]. In addition, biological cells are equipped with intelligent biomolecular engines (ATPase), which are demanded to generate biological fuel ATP, as shown in **Figure 1C** [3].

Micro-/nanomotors are micro-/nanoscale devices, which possess the ability to convert chemical energy into mechanical force or movement [4]. Evolution bestows

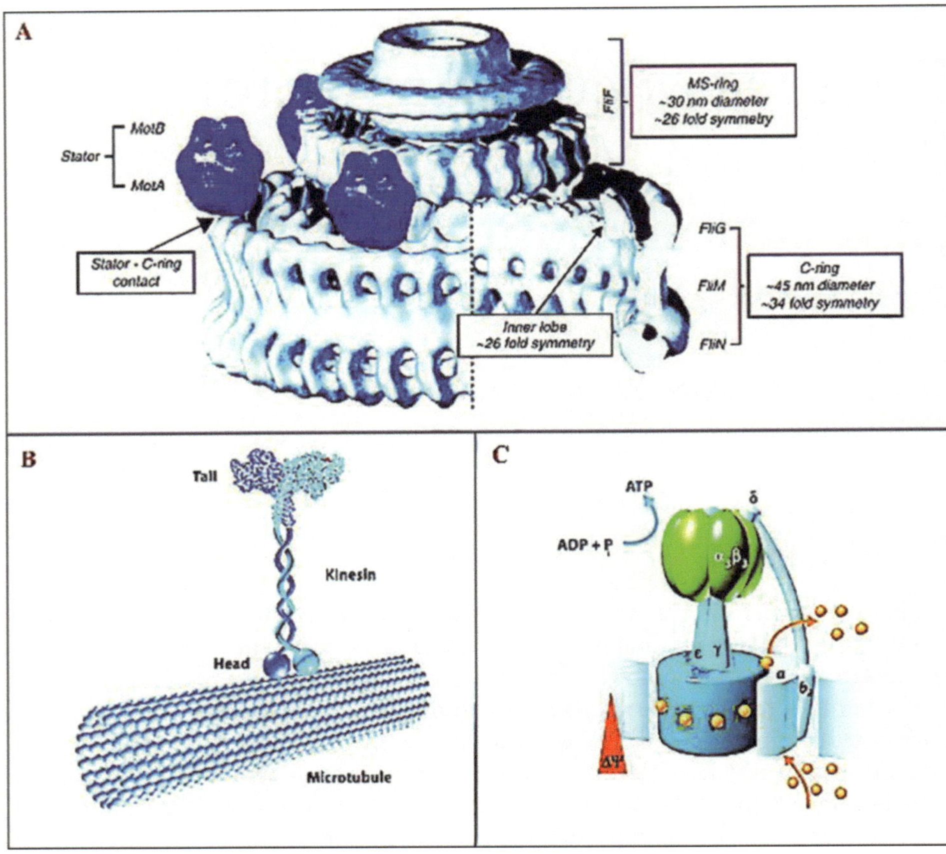

Figure 1.
(A) Schematic diagram of the architecture of the bacterial flagellar motor. (B) The twin heads of kinesin motor protein alternately bind to the microtubules so that the protein motor moves forward. (C) Schematic diagram of ATPase. Copyright 2012, Elsevier. Copyright 2004, IEEE Xplore Digital Library. Copyright 2000, The Royal society Publishing.

biomolecular protein motors with fascinating abilities to harness energy from living environments for autonomous motion *in vivo* as described above. Inspired by the fantasy of naturally occurring motor proteins, researchers paid great attentions into synthetic micro-/nanomotors in the past decades. In particular, led by pioneering contributions of Sen and Mallouk's team and Ozin's group, current work mainly focuses on the exploration of high-efficiency and high-speed synthetic micro-/nanomotors that are able to convert chemical energy into autonomous propulsion [4–6].

The research of synthetic self-propelled micro-/nanomotors has rapidly developed in last few decades [4–9]. Several advanced developments and excellent contributions had been made in this field. Although the bright future of this research area can be expected, some major existing challenges are still remained to be solved. The design, fabrication, and characterization of functional micro-/nanomotors require some innovative approaches and ideas to realize. Fabricating micro-/nanomotors with individual functional parts, smartly and precisely controlling motors are still extremely challenging. Hereby, a complete understanding of the physiochemical mechanism is necessary. To realize better control of micro-/nanomotors in the future, an industrial level of functional micro-/nanomachinery could be achieved. Despite of the significant development and advances in micro-/nanomotors, challenges are still remained to find specific relevant applications, such as biologically compatible fuels, etc.

Meanwhile, micro-/nanomotors have a wide variety of applications, including cargo towing, water cleaning, chemical sensors, biomedical applications, etc. [5, 7–9]. Advanced forms of micro-/nanomotors may stimulate and benefit other research. However, designing and powering micro-/nanomotors can be considered as a significant challenge in today's nanotechnology research. Hence, it is much beneficial for us to learn the state of the art of synthetic micro-/nanomotors and improve them in this research field. In this chapter, the reported work on the propulsion mechanisms, fabrication methods, propulsion controlling, and applications of synthetic self-propelled platinum-based micro-/nanomotors will be presented and discussed.

2. Propulsion mechanisms

Whitesides and his colleagues firstly reported the motion of a millimeter-scale object, which was composed of a piece of platinum (Pt)-coated porous glass filter mounted on a thin polydimethylsiloxane (PDMS) plate using a stainless steel pin, as shown in **Figure 2** [10]. The assembled object was immersed into hydrogen peroxide (H_2O_2) solution. Pt catalyzes H_2O_2 decomposition to generate oxygen (O_2) bubbles releasing from its surface, which reversely induced a recoil force to propel the object moving forward. This is the foundation in this research field. In micro-/nanoscale regime, only asymmetric particles can realize autonomous propulsion. Based on this, researchers focused on two aspects: shape and material compositions of micro-/nanomotors for breaking the symmetry, and various types of micro-/nanomotors were invented. Meanwhile, there were different mechanisms proposed to explain the propulsion phenomena, depending on the shapes (e.g., wires, rods, Janus spheres, and tubular jets) and material compositions of micro-/nanomotors [4–17].

2.1 Dielectrophoresis

In the past decades, a variety of micro-/nanomotors have been envisioned to explore the concept of self-electrophoresis propulsion, especially micro-/nanowires, rods, and Janus spheres. In self-electrophoresis, micro-/nanomotors produce a locally distributed electric field through chemical gradients and propel forward in

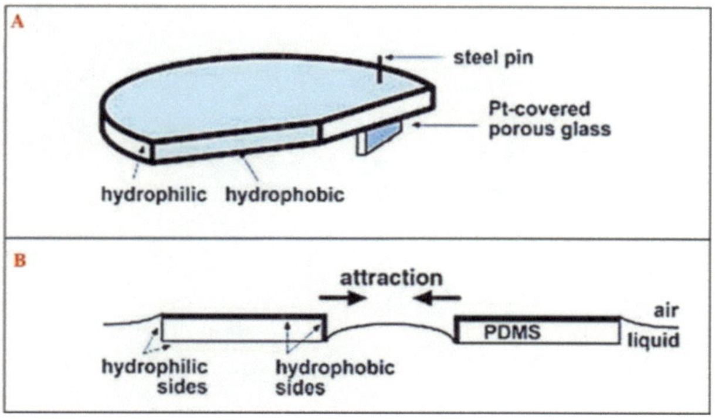

Figure 2.
(A) Schematic diagram of the propulsion of a millimeter-scale object. A thin plate (1–2 mm in thickness and 9 mm in diameter) was assembled from PDMS in a desired shape, and specified faces were rendered as hydrophilic by plasma oxidation. A 2 × 2 mm² piece of porous glass filter (one side covered with Pt) was mounted on the PDMS plate by using a stainless steel pin. (B) A diagram illustrating self-assembly by capillary interactions. Copyright 2002, Wiley Online Library.

respond to this self-produced electric field, and they do not respond to an external electric field.

Electrophoresis describes the movement of micro-/nanoscale particles in a fluid. In dielectrophoresis, if a micro-/nanoparticle would realize autonomous self-propulsion, the micro-/nanoparticle must contain at least two different metals and acts as a self-contained electro-chemical cell. For instance, a micro-/nanorod is composed of Pt and gold (Au), Au serves as the cathode and Pt is the anode, as shown in **Figure 3** [6].

The oxidation of H_2O_2 at the Pt side (anode) of the rod generates negatively charged electrons (e) and positively charged protons (H^+). The generated protons migrate along the double layer surrounding the rod and the electrons internally migrate along the rod from the Pt side to the Au side (cathode) of the rod; thus, H_2O_2 can be converted into water (H_2O) and O_2. The continuous chemical reactions result in a net flow of electrons from the anode to the cathode as well as the migration of protons to the cathode, producing a proton gradient along the axis of the rod. The electron flow produces a negatively charged rod that responses to the gradient, which propels the rod forward to the proton-rich environment previously occupied by Au [6].

In self-electrophoresis, the charged micro-/nanoparticles propel forward in a self-generated electric field resulting from an uneven distribution of ions. The velocity v of the particle is related to the self-produced electric field (E), zeta potential (ζ) of the particle, permittivity (ε), and viscosity (μ) of the medium, as shown below:

$$v = \frac{\zeta \, \varepsilon E}{\mu} \tag{1}$$

2.2 Diffusiophoresis

Self-diffusiophoresis is a propulsion phenomenon in which the movement of particles is induced by a concentration gradient of the reaction products. This propulsion mechanism is more commonly employed in spherical Janus micro-/nanomotors, as shown in **Figure 4** [11, 12]. In this system, the catalyst (Pt) is located at one side of the micro-/nanomotor; the reaction products (H_2O and O_2) preferentially accumulate at the site of the catalyst. Hence, a concentration gradient

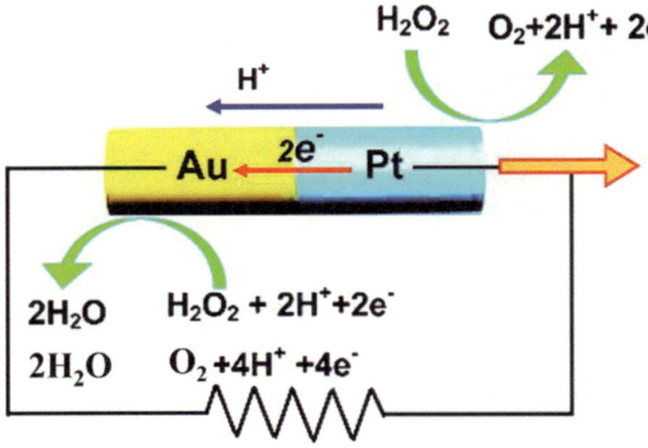

Figure 3.
Schematic representation of dielectrophoresis (bipolar electrochemical) mechanism for the propulsion of an Au-Pt micro-/nanomotor in the presence of H_2O_2. The mechanism involves an internal electron flow from one end to the other end of the motor, along with the migration of protons in the double layer surrounding the motor. Copyright 2006, ACS Publications.

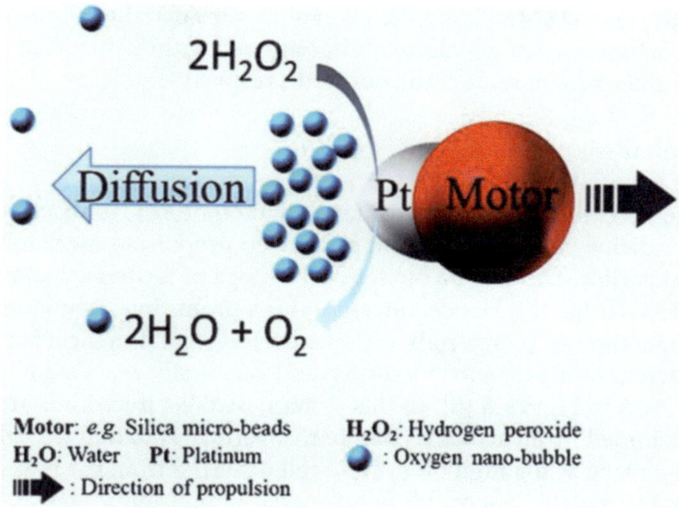

Figure 4.
Schematic diagram of a micro/nanomotor propelling under the diffusiophoresis propulsion mechanism. Copyright 2009, Wiley Online Library. Copyright 2011, Frontiers Journals of Higher Education Press.

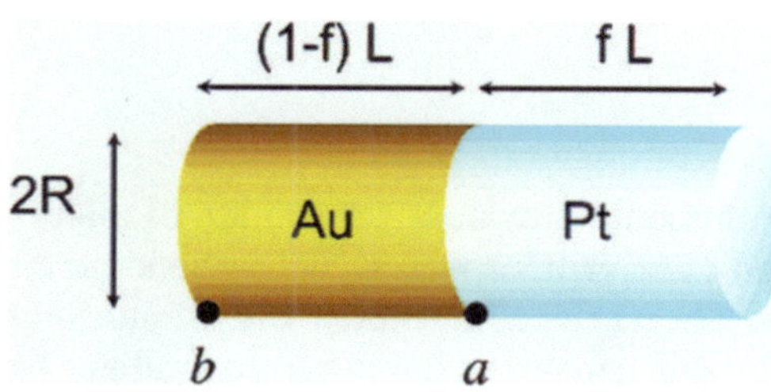

Figure 5.
Schematic illustration of a Pt-Au nanorod representing the dimensions used in the calculation of interfacial force. The parameter f is the length ratio of Pt. Copyright 2004, ACS Publications.

is produced along the surface of the micro-/nanomotor. As the reaction products reach a certain point, the local concentration is higher and the products start to diffuse away from the catalyst, which in turn produces a force leading to the movement of the micro-/nanomotor.

2.3 Interfacial tension

Surface tension gradient along an interface can result in an imbalanced force and further produce flow, which is well known as the "Marangoni effect." This motion mechanism has been employed to explain the propulsion of the micro-/nanomotors, and the model was firstly reported by Crespi, Mallouk, and Sen, as shown in **Figure 5** [13]. As H_2O_2 is decomposed at the Pt side of Au-Pt nanorods to produce H_2O and O_2, leading to an interfacial tension created near the surface of Pt is lower due to a larger quantity of O_2 generated. The surface tension difference between the Pt side and the Au side produces a force to thrust the micro-/nanomotor propelling forward.

In this model, it is proposed that the velocity v of micro-/nanomotors is linearly proportional to the surface tension of the solution, as shown in the following equation:

$$v = \frac{SR^2\gamma}{\mu DL} \propto k\gamma \tag{2}$$

where S, R, γ, μ, D, L, and k are the O_2 evolution rate, radius of micro-/nanomotor, surface tension of solution, viscosity of solution, diffusion coefficient and length of micro-/nanomotor, and constant, respectively.

2.4 Acoustophoresis

One of the potential applications of micro-/nanomotors is to be used in diagnostics and biomedicine; there is a demand to develop propulsion mechanisms which can be biocompatible. Ultrasound operates in a range of frequency above 20 kHz, which is biocompatible [5]. Hence, ultrasound is a promising technique for propelling micro-/nanomotors (wires/rods and tubular jets) in biomedical applications. The micro-/nanomotor propelled by ultrasound was firstly reported by Sen's group in 2012, as shown in **Figure 6** [5]. In this system, metallic microrods are suspended in water surrounded by an acoustic chamber. A vertical standing wave levitates the microrods to a plane at the midpoint of the cell resulting from the lowest pressure. In that plane, the metallic microrods behave axial propulsion at speeds up to 200 μm/s in water. The microrods also assemble into patterns in the nodal plane resulting from nodes and antinodes among the plane. The composition of the microrods was observed to effectively affect their propulsion, with only metallic microparticles demonstrating fast axial motion.

2.5 Thermophoresis

Temperature gradient could introduce the motion of micro-/nanoparticles (micro-/nanowires, rods, and Janus spheres). This phenomenon is called thermophoresis or the "Soret effect." Recently, the propulsion of micro-/nanomotors induced by self-generated temperature gradient has been investigated, as shown in **Figure 7**.

Jiang et al. studied self-thermophoresis at the single particle level (**Figure 7A**) in 2010 [14]. Janus silica (SiO_2) microspheres half-coated with Au were irradiated in water by using a defocused laser beam at 1064 nm. Absorption of laser by the thin Au layer generated heat, resulting in a local temperature gradient (~2 K across particles) which produced thermophoresis. After that, Baraban et al. applied an

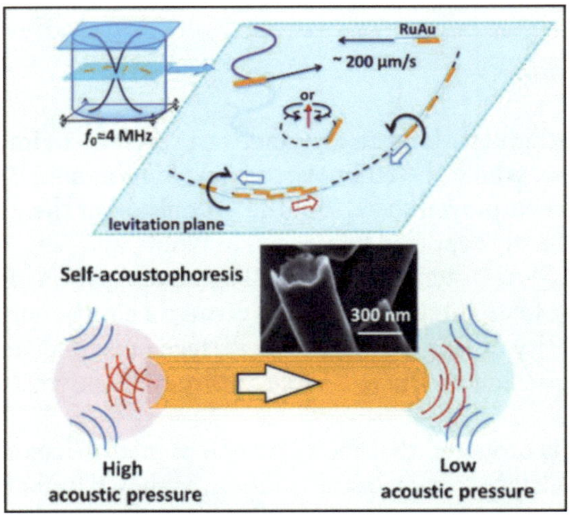

Figure 6.
Self-acoustophoresis mechanism: asymmetrically shaped metallic microrods are triggered by an ultrasonic standing wave at MHz frequency. Copyright 2012, ACS Publcations.

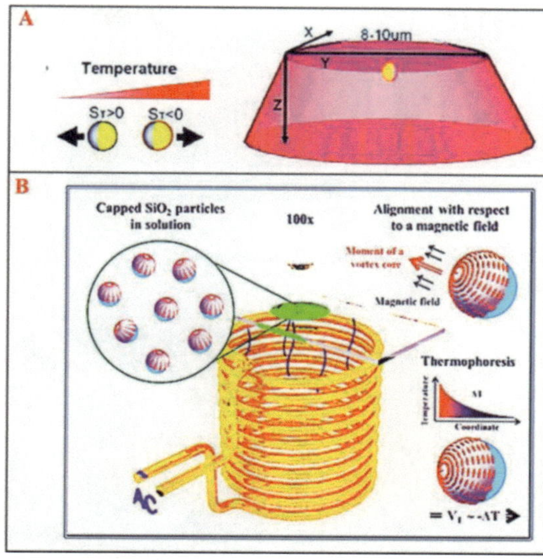

Figure 7.
Self-thermophoretic microparticles. (A) Au-capped SiO_2 microspheres undergoing autonomous propulsion resulting from the "Soret effect" in a defocused laser beam. (B) Permalloy-capped SiO_2 particles moving by self-thermophoresis in an AC magnetic field. Copyright 2010, APS Publications. Copyright 2012, ACS Publications.

alternating current (AC) magnetic field to heat up and down permalloy-capped SiO_2 particles in solution and viewed autonomous movement, as shown in **Figure 7B** [15].

Short heat pulses have also been applied to modulate the speed of micro-/nanomotors. Au-Pt nanowires subjected to elevated temperatures were observed to propel forward extremely faster than those at room temperature. For instance, an average speed of 45 μm/s was obtained for nanowires at 65°C compared with 14 μm/s at 25°C. The speed increasing could be resulted from the reduction of the solution's viscosity and the temperature dependency of the electrochemical process. The utilization of heat pulses is an extremely reversible process, after incorporated with magnetic steering, which will enable a more advanced spatial and temporal control, with the capability to modulate both the direction and the speed.

2.6 Bubble propulsion

Bubble propulsion is possibly the most commonly studied mechanism in the field of micro-/nanomotors, which can be suitable for motors with any shapes as long as the motors are decorated with catalysts. The motion of the motors is produced by the releasing of micro-sized bubbles from the decomposition of fuel catalyzed by the catalyst. The most extensively studied instances of bubble-propelled micro-/nanomotors are those decorated with Pt as a catalyst to decompose H_2O_2 into H_2O and O_2 bubbles, as shown in **Figure 8** [16, 17].

Bubble propulsion originates from the spontaneous decomposition of a fuel stimulated by a catalyst into micron-sized gas bubbles, whose detachments from the micro-/nanomotor's surface produce a recoil force to thrust the movement of the motors in the direction away from the catalyst. These motors, whose size ranging from a few micrometers to hundreds of micrometers, can realize powerful movement with considerable speed. In bubble-propelled micro-/nanojets, Solovev and his colleagues reported that the speed of the micro-/nanojets was related to the bubble's detachment frequency and radius, with the deviations at large values

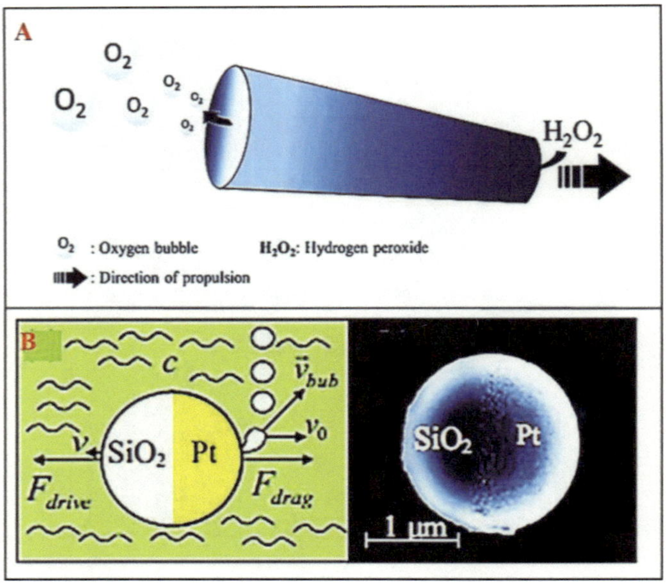

Figure 8.
Examples of bubble propelled micromotors. (A) Schematic illustration of a tubular micro-/nanomotor's movement by the bubble propulsion mechanism. (B) Janus Pt-SiO$_2$ spheres moving in H$_2$O$_2$. Copyright 2014, RSC Publications. Copyright 2009, AIP Publications.

resulted from the potential collisions between bubbles, which diminished the distance traveled by the bubbles within the tube and limits the detaching stage. They concluded that the dynamics of micro-/nanojets is influenced by their shapes, the fuel composition, and the viscosity of the medium.

3. Fabrication methods

Artificial micro-/nanomotors could offer a significant advance in the field of miniaturized devices. However, a major challenge in artificial micro-/nanomotor field is the synthesis of such tiny devices with high quality and reliability. Relying on the intended applications, different synthesis techniques must be taken into account, since each specifically shaped micro-/nanomotor demands a specialized synthesis technique as well as defines the desired propulsion mechanism. The rapid developing of nanotechnology has resulted in various techniques and strategies for the synthesis of micro- and nanoscale motors. The purpose of this section is to present versatile schemes to the synthesis of micro-/nanomotors. The synthesis strategies and the factors that ought to be considered in the design of micro-/nanomotors are including the shapes, compositions, and distributions of materials, and functionalization. By highlighting the progresses that have been achieved in the synthesis of artificial micro-/nanomotors over the last decades, we intend to present the challenges and opportunities facing synthesis and put forward perspectives for the upgrowth of new methods.

3.1 Electrochemical deposition

Electrochemical deposition is a process that applies external electric current to grow materials and enables the growth of arbitrary three-dimensional (3D) shapes with distinct materials varying from metals to polymers, resulting in the widespread applications of this growth strategy in nanotechnology. The process can be conducted without the requirements of expensive instruments and harsh working environments. Therefore, micro-/nanostructures with diverse dimensions can be grown using this method, especially the micro-/nanowire, micro-/nanorod, and tubular micro-/nanoengines.

Template-assisted electrochemical deposition, as shown in **Figure 9**, utilizes the pores of a membrane template to grow the required wires and tubes comprising of different materials [18]. Each pore of the template functions as a reactor in which the desired structure is synthesized. Membrane templates commonly used for the synthesis of micro-/nanomotors are track-etched polycarbonate (PC) membranes and porous alumina (AAO) membranes. Relying on the properties of the material and the chemistry of the pore wall, the micro-/nanomotors can be either hollow or solid. Membrane template-assisted electrodeposition provides a relatively low-cost and powerful approach for synthesizing micro-/nanowires, micro-/nanorods, and tubular micro-/nanoengines.

For membrane template-assisted electrodeposition, a layer of Au or silver (Ag) is firstly coated on one side of the membrane by physical vapor deposition (PVD) to play a role of the working electrode. Afterwards, the membrane is assembled in a Teflon plating cell with flat aluminum (Al) foil located against the metal layer to work as a conductive contact for subsequent electrodeposition. Commonly, a sacrificial layer of Ag or copper (Cu) is firstly grown, followed by sequential growth of different required metals. The Ag or Au backing and the sacrificial layer are etched away by chemical etchant or are removed by mechanical polishing. By removing the alumina (Al_2O_3) membrane in sodium hydroxide (NaOH) solution, the nanowires or nanotubes can be released and obtained after successive rinsing and centrifugation.

The Wang's and Pumera's research groups combined electrodeposition widely applied in the growth of nanowire-based micro-/nanomotors with the bubble-propelled tubular micro-/nanojets, as shown in **Figure 10** [19–24]. Two geometries (either cylindrical or conical) can be obtained, entirely relying on the type and geometry of porous template (either PC or AAO). In addition to such metals, the incorporation of polyethylenedioxythiophene (PEDOT), polyaniline (PANI), and polypyrrole (PPy) polymers with Pt generates catalytic microjets. On the other hand, the wall can be incorporated with molecularly imprinted polymers (MIPs) so that alternative recognition cavities can be implemented for the selective separation of biomolecules.

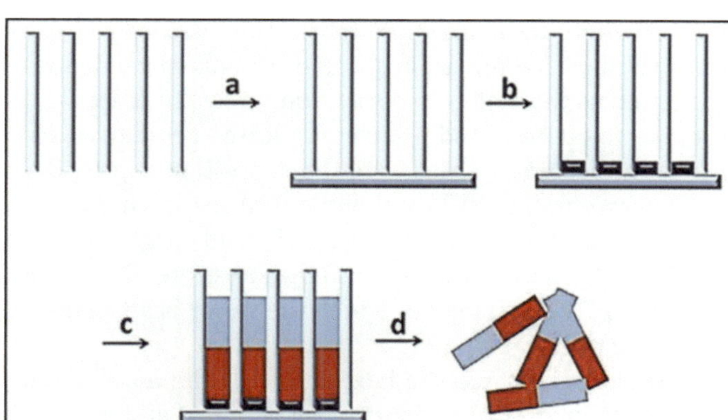

Figure 9.
Membrane template-assisted electrochemical deposition of micro-/nanomotors: (a) coating of Au or Ag backing on the membrane template, (b) electrochemical deposition of the sacrificial layer, (c) sequential electrochemical deposition of desired elements, and (d) removal of the backing and sacrificial layer and dissolution of the membrane. Copyright 2015, ACS Publications.

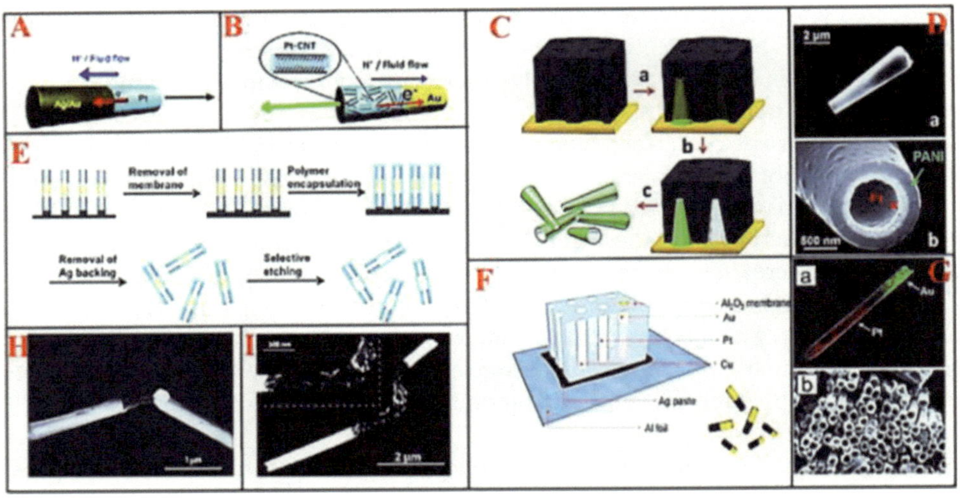

Figure 10.
Examples of micro-/nanomotors grown by template-assisted electrochemical deposition. (A) Electrodeposited Ag-Au/Pt and (B) Au/Pt-CNT nanomotors in H_2O_2. (C, D) Polycarbonate membrane-assisted growth and SEM images of conical PANI/Pt microtubes, respectively. (E) Growth procedures of flexible metallic nanowires with polyelectrolyte hinges after membrane template electrodeposition. (F, G) Anodized AAO membrane-assisted growth and SEM (scanning electron microscopy) images of segmented microtubes, respectively. (H) SEM image of a hinged nanowire. (I) SEM image of a Au/Agflex/Ni nanomotor with flexible central Ag segment. Copyright 2008, Wiley Online Library. Copyright 2008, ACS Publications. Copyright 2011, ACS Publications. Copyright 2007, Nature Publishing Group. Copyright 2013, RSC Publications. Copyright 2010, ACS Publications.

3.2 Physical vapor deposition

PVD is a vaporization deposition process for coating thin layers of materials. The material from a solid target is firstly vapored by a gaseous plasma or a high-temperature vacuum. Afterwards, the vapor is transferred to the surface of the substrate in vacuum or partial vacuum. Finally, it is condensed to produce thin films. The two most common kinds of PVD procedures are electron beam evaporation and sputtering. Electron beam evaporation is a process that generates an electron beam to evaporate atoms from the target into the gaseous phase, whereas sputtering creates vapor through bombardment of the target by ionized gas, typically argon (Ar). In both strategies, the produced vapor phase is subsequently condensed onto the surface of the substrate.

PVD has been shown to be an effective fabrication method in micro-/nanomotors. Compared with template-assisted electrochemical deposition, PVD has some advantages, such as the ability to coat a wide range of materials, less fabrication processes, easier to operate, and more complicated geometries of micro-/nanomotors can be fabricated. According to the deposition angles, there are two categories of PVD: conventional PVD and dynamical shadowing growth (DAG). In conventional PVD, the substrate is placed parallel to the target and the vapored metal flux is condensed almost vertically onto the substrate. DAG or glancing angle deposition (GLAD) is a PVD strategy in which the vapor is deposited onto a substrate at an oblique angle.

Posner and co-workers reported the fabrication of a bimetallic spherical motor depending on the electrophoretic mechanism for motion, as shown in **Figure 11A** [25]. Firstly, the microspheres were half-deposited with Au by sputter machine. Afterwards, they were resuspended in water and coated again with Au in random orientation, which was repeated seven or eight times until the whole surface was fully deposited with Au. Finally, the Au-deposited spheres were half-deposited with Pt, which produced bimetallic spherical Janus micromotors that are able to propel

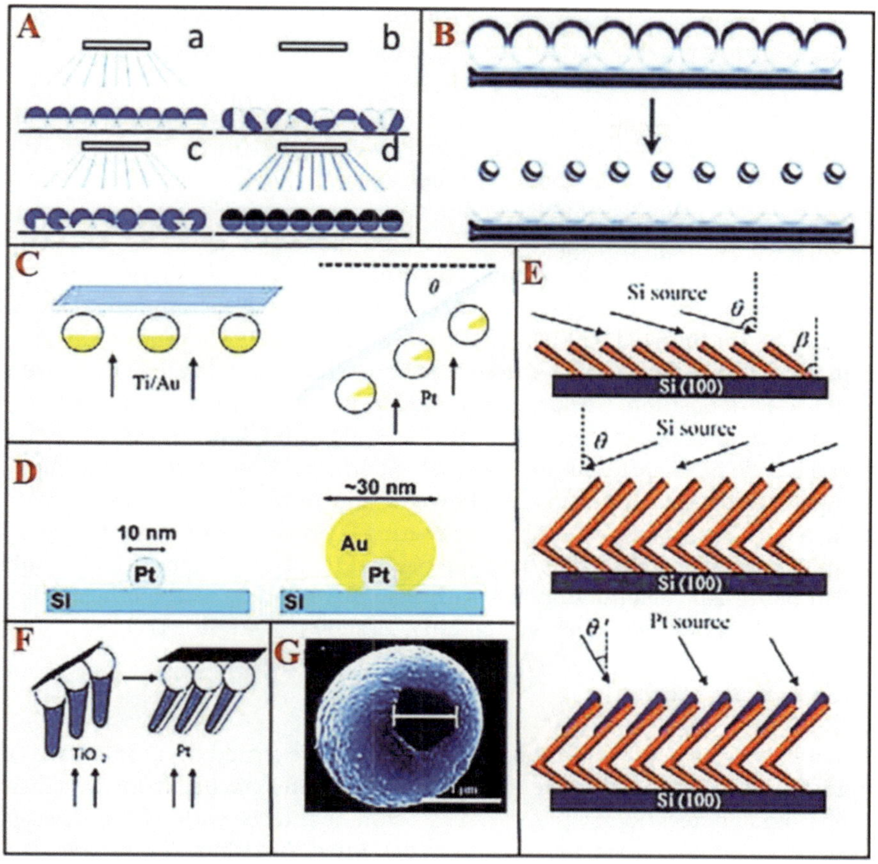

Figure 11.
(A) Schematic representation of the synthesis of bimetallic Janus micromotors by conventional PVD. (B) Synthesis of sphere dimers via thermal annealing. (C) Fabrication of asymmetric Pt/Au-coated catalytic micromotors by GLAD. (D) Fabrication of electrophoretic Pt-Au Janus nanoparticles by GLAD. (E) Synthesis procedures of L-shaped Si/Pt nanorod motors by GLAD. (F) Fabrication of catalytic micromotor comprising of a spherical SiO_2 colloid with a TiO_2 arm deposited asymmetrically with Pt. (G) SEM image of a Pt-Ag-Au shell micromotor synthesized by GLAD. Copyright 2010, ACS Publications. Copyright 2010, Wiley Online Library. Copyright 2010, AIP Publications. Copyright 2014, ACS Publications. Copyright 2007, ACS Publications. Copyright 2004, ACS Publications. Copyright 2013, RSC Publications.

forward at velocities comparable to their nanowire counterparts. On the basis of sphere templates, PVD can be employed to synthesize not only spherical Janus motors but also versatile motors with various geometries. Valadares and co-workers studied a catalytic dimer comprising of a Pt half-sphere and a SiO_2 sphere. The spheres were firstly deposited with a bilayer of Cr/Pt by using sputter machine, followed by an annealing process, during which the metallic half-shell formed a Pt particle combined with the SiO_2 sphere, as shown in **Figure 11B** [26].

Relying on the substrate rotation during the deposition of an incident vapor and the self-shadowing effect, GLAD offers an easier way to synthesize Janus micro-/nanomotors with complicated geometries. Zhao and his colleagues studied the asymmetric Pt/Au-deposited catalytic micromotors synthesized by GLAD. To get an asymmetric bimetallic deposition, a SiO_2 microbeads-coated substrate was rotated to a polar angle after coating of an adhesive titanium (Ti) layer and an Au layer, the subsequent Pt coating left some of the Au layer exposed, as shown in **Figure 11C** [27]. The propulsion behavior could be regulated by changing the exposed area of the Au layer. Lee and co-workers synthesized a 30 nm Pt/Au Janus nanomotors by GLAD in which Au under fast substrate rotation was deposited onto an array of Pt nanoparticles generated by block copolymer micelle lithography, as shown in **Figure 11D** [28]. Both of the bimetallic Janus motors depend on the self-electrophoresis mechanism for propulsion.

He and his colleagues demonstrated the synthesis of rotary silicon/Pt (Si/Pt) nanorods, Si/Ag nanorods, and L-shaped Si/Pt where they firstly applied GLAD to synthesize the Si nanorod backbone and then asymmetrically deposited a Pt or Ag layer on one side of the nanorod backbone with a geometric shadowing effect. The L-shaped backbone was fabricated by a high speed of azimuthal rotation of the substrate in the middle of oblique angle deposition, as shown in **Figure 11E** [29]. By monitoring the substrate rotation and the deposition angle, complex rolling Si/Ag springs can be synthesized. Gibbs and Zhao reported the rotary propulsion of a micromotor comprising of a SiO_2 microbead and a titanium dioxide (TiO_2) arm with asymmetric Pt deposition. The arms of the micromotors were synthesized on the closely packed microbeads at oblique angles. As such, the Pt is subsequently coated only on one side of the arms at no angle, which offers the asymmetric placement of the catalyst critical for propulsion, as shown in **Figure 11F** [13]. With substrate rotation and oblique vapor direction, the coating layer can cover a much bigger area of the sphere templates than that by using conventional vapor deposition. A bubble-propelled Pt-Ag-Au shell micromotor with a smaller opening was synthesized by GALD and subsequent wet chemical etching, as shown in **Figure 11G** [30].

3.3 Rolled-up nanotech

By combining an engineered strain gradient with the coated thin membranes, the membranes are able to roll into the required shapes when detached from the substrate. The rolled-up nanotechnology pioneered by Schmidt and co-workers applies strain engineering to form micro-/nanotubes from deposited thin films. A prestressed nanomembrane is coated onto a photoresist sacrificial layer patterned by photolithography, which is able to be alternatively etched by acetone. GLAD deposition is applied to guarantee accurate positioning and tube integration on a single chip. A proper control of the deposition rate and the substrate temperature, as well as the stress evolution during coating, creates the strain gradient desired for the rolling process.

The coated nanomembrane forms into a microtube once detached from the substrate by the dissolution of the sacrificial layer, as shown in **Figure 12A** [31, 32]. To avoid collapse of the rolled-up nanomembranes, the critical drying point is required to dry the synthesized microtubes. Microtubes with distinct opening diameters varying from 1 to 30 μm can be synthesized by modulating the built-in strain and the thickness of the nanomembranes. The lengths of the microtubes are in the range of scores of micrometers. Catalysts such as Pt consist of the inner wall of the microtubes by simply being coated onto the top layer of the nanomembranes. The wrinkle orientation of the detached membranes is defined by the different etching rates along the crystal axis and the crystal structure of the sacrificial layer.

Due to the high cost and the complex fabrication procedures of the rolled-up technique, considerable efforts have been devoted to simplifying the rolled-up procedures and decreasing its cost. Microtubes with outer layers of graphene oxide (GO) were synthesized by coating metal layers on GO nanosheets, as shown in **Figure 12B** [33]. Microscrolls with GO on the outer side and Pt at the inner surface were spontaneously synthesized upon sonication, resulting from material strain and weak bonding between GO layers. The diameter can be changed by modulating the thickness of the coated metal layers. A similar fabrication process of tubular microengines was demonstrated using accessible and low-cost fruit cells as support for the metallic layers. The tissue-based microengines can demonstrate extremely efficient bubble motion in the presence of H_2O_2. Zhao and his colleagues reported the synthesis of Pt microtubes by alternative dissolution of the poly(methyl

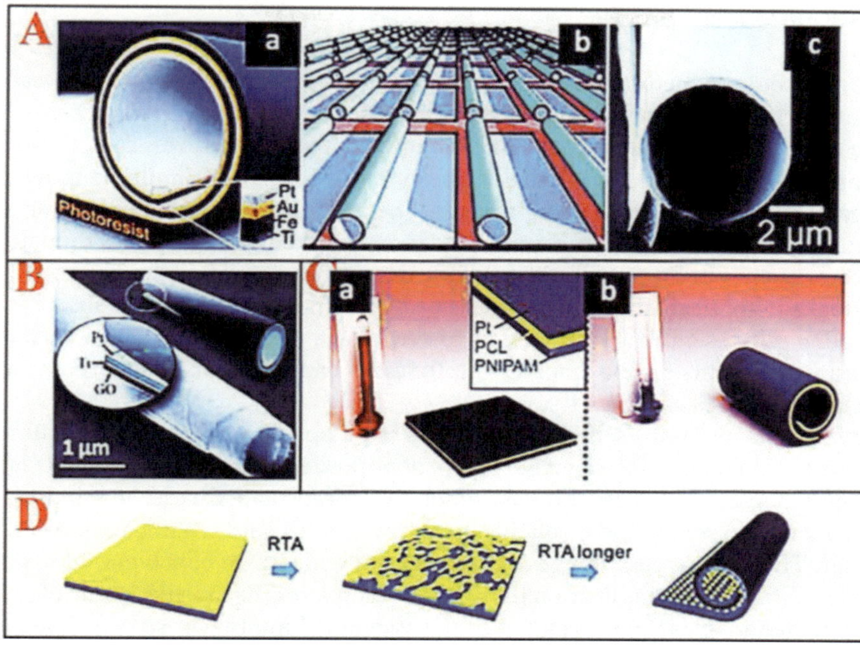

Figure 12.
Rolled-up nanotech. (A) Rolling-up of nanomembranes patterned with photoresist: (a, b) schematic illustration of a rolled-up microtube comprising of Pt/Au/Fe/Ti multilayers on a sacrificial photoresist layer and an array of rolled-up microtubes, respectively; (c) SEM image of a rolled-up microtube. (B) Rolled-up microtubes with GO as an outside layer. (C) Reversible rolling and unrolling of thermoresponsive polymeric Pt microtubes. (D) Particle-aided rolling process of nanomembrane upon a thermal dewetting treatment. Copyright 2009, Wiley Online Library. Copyright 2010, Wiley Online Library. Copyright 2012, ACS Publications. Copyright 2014, Wiley Online Library. Copyright 2013, Wiley Online Library.

methacrylate) (PMMA) sacrificial layer beneath the sputtered Pt layer or by H_2O_2-assisted lift-off of the Pt layer coated directly on a glass wafer. A transmission electron microscopy (TEM) grid template was applied to synthesize microtubes with a relatively uniform size. Despite the low cost and simple methods described above provide great possibilities for large-scale yield of microtubes, a major issue of these methods is the lack of morphology and accurate control of the size of the rolled-up microtubes. On the premise of simplifying the processes, future efforts should be devoted to the better manipulation of the rolled-up technique.

Magdanz and his colleagues reported a flexible thermoresponsive polymeric microjet resulting from the reversible folding/unfolding of the polymer at decreased and elevated temperatures (**Figure 12C**) [34]. Cooling of the Pt/polymer layers results in folding of the films and synthesis of microtubes in diameter 30 μm with a Pt inner layer as a catalyst, while warming leads to the unfolding of the microtubes. Hence, the rolling and unrolling procedures of the microtubes could be conducted reversibly by changing the temperature of the solution to start and stop the propulsion of the microtubes. The diameters of the microtubes synthesized by the rolled-up nanotech are all in the range of microscale.

Li and his partners took advantage of the surface tension of nanodroplets as well as the intrinsic strain relaxation in the nanomembranes to reduce the diameters of the rolled-up tubes to hundreds of nanometers. A layer of Pt was coated onto a prestrained bilayer of SiO_2/TiO_2 or Si/Cr on a sacrificial PMMA layer, as shown in **Figure 12D** [35]. In this treatment, rapid thermal process (RTP) was applied to stretch the Pt layer to the isolated islands and the nanodroplets brought considerable surface tension for rolling. On the other hand, the removal of PMMA resulted in the detachment of the nanomembranes. The synthesized microtubes exhibit higher velocities compared with those with a smoother Pt surface.

3.4 Advanced assembling

The aforementioned fabrication methods: template-assisted electrochemical deposition, PVD, and rolled-up nanotech, are effective approaches for synthesizing micro-/nanomotors. However, to achieve more complex structures, the assembly technique must be developed. The construction of devices with multiple individual tiny parts is an extremely challenging task. The assembling approach plays an essential role in micro-/nanofabrication. It is a technique that combines miniaturized components to form a required device. The unique properties of the assembling make it applicable for the synthesis of micro-/nanomotors. Not only employing self-assembling of materials to synthesize the required devices, but also the desired elements can be embedded into micro-/nanomotors by incorporating them into materials as well.

Layer-by-layer (LbL) self-assembling is a nanofabrication strategy for multilayer formation by coating selective layers of oppositely charged materials. It is an easy-operation and low-cost process, which can encapsulate diverse materials, such as tiny inorganic compounds, colloids, macromolecules, and organic molecules together. The LbL process can be applicable for a wide range of solvent-accessible surfaces, allowing the application of different templates. Encapsulation of Pt nanoparticles enables the movement of the assembled multilayer structure to be driven in H_2O_2 solution. Taking advantages of simplicity, versatility, and low cost, the LbL assembling, primarily employing the electrostatic interaction between oppositely charged species, has been widely employed to synthesize various multilayer materials.

He's group firstly reported the combination of a colloid template-assisted LbL assembling with a microcontact printing method to synthesize platinum nanoparticles (Pt NPs) asymmetrically coated autonomous Janus micromotors, as shown in **Figure 13A** [36]. The SiO_2 particles as templates were selectively dispersed into positively charged polyallylamine hydrochloride (PAH) solution and negatively charged polystyrene sulfonate (PSS) solution to form one polyelectrolyte

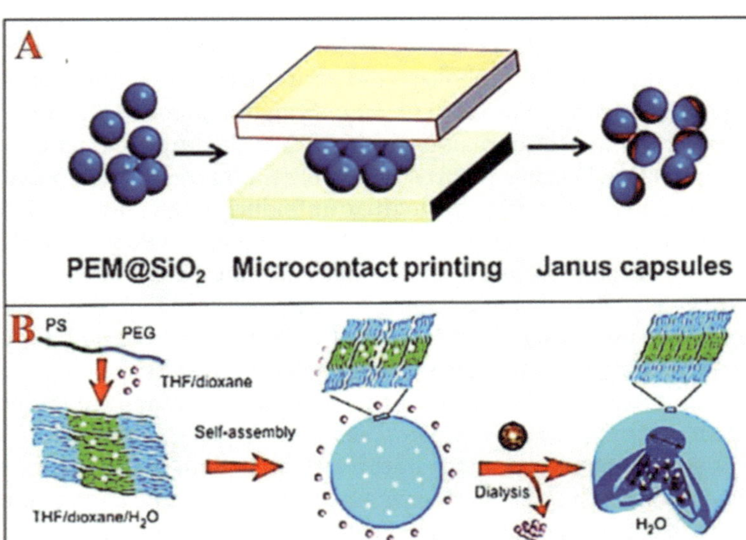

Figure 13.
Schematic diagram of various types of controllable self-assembled micro-/nanomotors. (A) Synthesis process of Pt NPs-functionalized Janus capsule motors. (B) Selective and controlled encapsulation of Pt NPs inside artificial stomatocytes during shape transformation. Copyright 2012, ACS Publications. Copyright 2012, Nature Publishing group.

bilayer. After five bilayers were coated, the formed particles were spread onto a glass wafer to form a monolayer that was then printed with a PDMS stamp loaded with a drop of dendritic Pt NPs suspension. The Pt NPs partially loaded hollow Janus micromotors were collected after the removal of the templates by using hydrofluoric (HF) acid. Polymer stomatocytes that are able to entrap Pt NPs into their nanocavities were reported by Wilson and collaborators, as shown in **Figure 13B** [37]. This approach applied the controlled transformation of spherical polymersomes into stomatocyte structures. Entrapment of the Pt NPs was achieved by adding nanoparticles to solvent-swollen polymersomes during the transformation process. H_2O_2 can get in touch with the catalytic Pt NPs by controlling the entrance of the stomatocytes to produce driving force for the directional motion of the stomatocytes.

4. Controlling methods

The control of micro-/nanomotors is essential to meet various requirements in practical applications. The precise propulsion control of micro-/nanomotors is leading to advances in practical applications, and thus it is quite critical to put forward the controlling strategies for micro-/nanomotors. In the past years, scientists have realized the propulsion control of micro-/nanomotors by using different methods as reported below.

4.1 Magnetic control

External magnetic field is the most common control source employed to direct and guide the micro-/nanomotors. The predetermined trajectory of micro-/nanomotors can be realized by incorporating a paramagnetic or ferromagnetic part that can be magnetized by the magnetic field. Relying on the shapes of micro-/nanomotors, the magnetic part can be introduced by either electrodeposition or PVD. The appropriately used magnetic material candidates in micro-/nanomotors are nickel (Ni) and iron (Fe).

Wang et al. reported a multifunctional nanomotor with three segments (Au-Ni-Au), which was thrust by ultrasound and steered by the magnetic field. A concavity was also decorated at the end of the Au segment by the sphere lithography process to realize asymmetric geometry. The interaction between the magnetic field and the middle magnetic Ni segment produced a predefined and controllable movement of the nanomotor. The Ni segment was used to load and deliver magnetic particles along a predetermined route as well, as shown in **Figure 14A** [38].

Magnetic orientation has proved to be extremely effective for achieving the required directionality of the self-assembled motors. Sputtering a layer of magnetic material on one side of the motors is widely applied in Janus capsule motors. The catalase-functionalized Janus capsule motor was coated with a layer of 5-nm-thick Ni before the deposition of Au. Such biocatalytic Janus capsule motors were capable of swimming in cellular media in the presence of H_2O_2 fuel and were steered by the applied magnetic field toward the targeted HeLa cells, as shown in **Figure 14B** [39]. It should be mentioned that the magnetic field is exclusively employed to steer the propulsion directionality of motors and is not strong enough to initiate the propulsion of motors by magnetic attraction.

For electrodeposited nanowire/-rod and micro-/nanotube, a Ni part can be incorporated into the structure by using electrodeposition. A self-propelled segmented Pt/Ni/Au/Ni/Au nanowire was taken as one of the earliest examples, as shown in **Figure 14C** [40]. The nanowire was magnetized transversely rather than

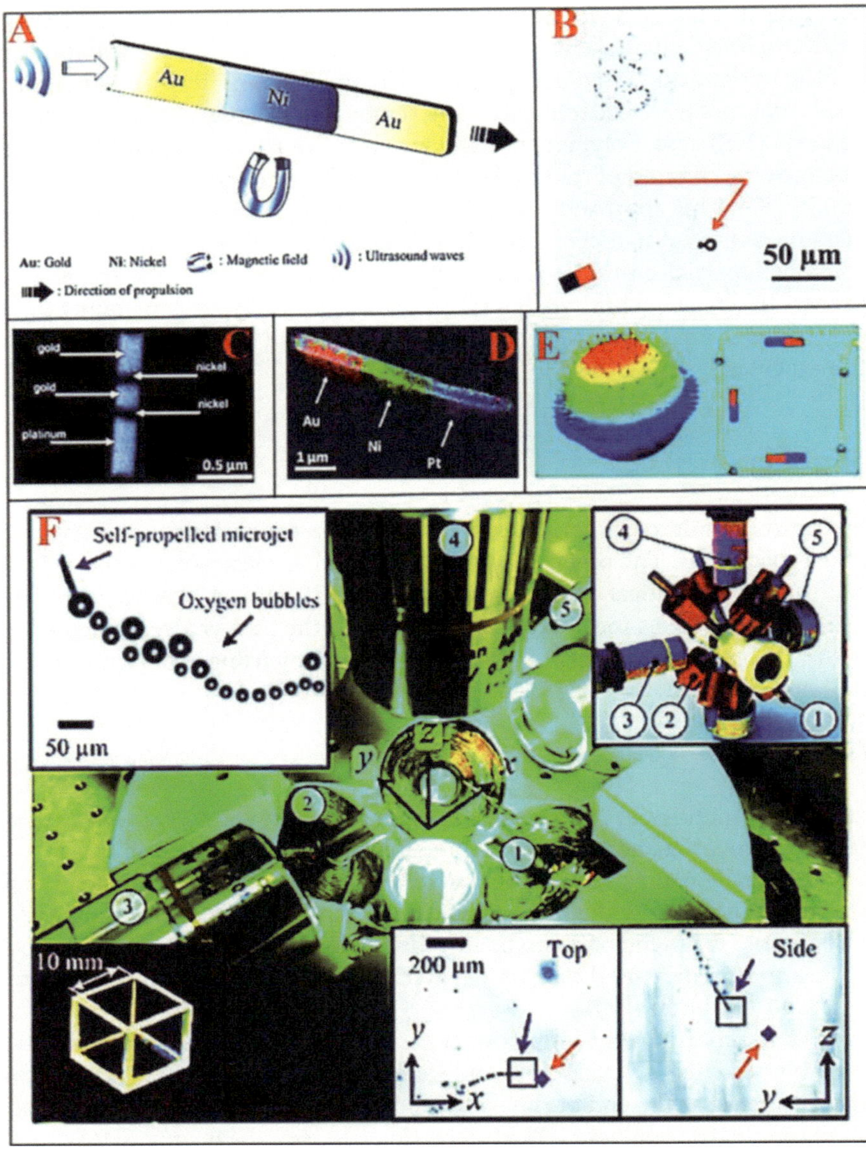

Figure 14.
Micro-/nanomotors controlled by the magnetic field. (A) Schematic diagram of an Au–Ni–Au metal alloy propelled by ultrasound and steered by the magnetic field. (B) Magnetically steered movement of Janus capsule motors toward targeted HeLa cell sheets. (C) SEM image of Pt/Ni/Au/Ni/Au nanowire. (D) SEM/energy-dispersive X-ray (EDX) elemental analysis of Au/Ni/Pt nanotube. (E) Scheme representing the magnetic steering of Janus micromotors. (F) Remote control of micro/nanojets by magnetic field. Copyright 2013, ACS Publications. Copyright 2014, ACS Publications. Copyright 2005, Wiley Online Library. Copyright 2013, ACS Publications. Copyright 2012, ACS Publications. Copyright 2013, AIP Publications.

longitudinally, resulting from the scale of the electrodeposited Ni part was smaller than the diameter of the wire. Magnetized nanowire can orient its net magnetic moment parallel to an external magnetic field, resulting in precise steering by manipulating the orientation of the magnetic field. Experimental results proved that the magnetic field could only direct the nanowires without changing their speed. The steered propulsion of self-propelled Au/Ni/Au/Pt-CNT nanorod as well as the delivery of magnetic microbead cargoes by it in microchannel networks was reported by Burdick et al. [20].

Magnetic steering of the electrodeposited microtube could also be achieved by additional electro-deposition of Ni. For conical microtube, Ni is grown to cover the whole inner surface of the microtube before electrodeposition of Pt, resulting in the microtube is magnetized along the tube axis. A simplified Ni/Pt alloy inner layer obtained by the co-deposition of a Ni/Pt layer can display both magnetic and catalytic properties. Unfortunately, the speed of microtube with a simplified Ni/Pt alloy inner layer in diluted H_2O_2 is hugely decreased because of the reduced catalytic area. For striped nanotube with different elements placed longitudinally, it can also be magnetized longitudinally due to the Ni portion has a larger dimension along the axis of the tube, demonstrating behavior similar to that of magnetotactic bacteria, as shown in **Figure 14D** [41].

Regarding rolled-up microtube, an additional Fe layer can be integrated into the microtube during the deposition process to realize the magnetic control. The longitudinally magnetized rolled-up microtube can monitor the direction of an external field and orient itself accordingly. A magnetized Fe-contained microtube was studied to be able to selectively pick up and deliver paramagnetic beads in the absence of an external magnetic field. The ability of a Pt Janus particle to deliver cargo has been reported by Sanchez and fellows. To realize the better control of the propulsion of catalytic Janus motors and the cargo delivery process, magnetic caps consisting of (Co/Pt) multilayers were incorporated into the structure by PVD. The magnetic caps were envisioned to align the magnetic moment along the main symmetric axis of the cap, enabling direct manipulation of the Janus motor as well as superparamagnetic cargoes delivery by using an external magnetic field, as shown in **Figure 14E** [42]. Precise control of magnetic Janus particles is further reported by sorting beads between the channels in microchip devices.

The individual control of microjet in a closed-loop manner and 3D propulsion control were the next steps to be considered. Recently, Misra's and Sanchez's research groups presented the accurately closed-loop control of microjet [43]. The authors reported precisely point-to-point closed-loop control by applying weak magnetic fields (2 mT). Another study demonstrated precise control when a flow was employed against and along the propulsion direction of the microjet. An electromagnetic setup consisting of two sets of orthogonal arrays of electromagnetic coils with a Fe core in conjunction with two microscopic systems was employed to guide the movement of microjet in 3D space, as shown in **Figure 14F** [43]. Microjet overcomes vertical forces, such as vertical flow buoyancy forces, and interaction forces with O_2 bubbles, and thus it is able to drive downwards and swim upwards relative to reference positions.

4.2 Acoustic control

Ultrasound not only provides energy for the motion of micro-/nanomotors but also offers an alternative manner for controlling self-propelled motors. Using ultrasound to guide micro-/nanomotors and as rapid "stop/go" switching of micro-/nanomotors in respond to "on/off" of ultrasound are investigated.

The movement direction of nanomotors can be reversed by varying the power of the ultrasound field. Fast and reversible transitions between aggregated and free-moving states of nanomotors in H_2O_2 were obtained in response to switching between on and off ultrasound states, as shown in **Figure 15A** [44]. The generation of bubbles can be disrupted by the ultrasound field. Wang et al. demonstrated the reversible control of the propulsion of PEDOT/Ni/Pt microengines by changing the applied voltage of the external transducer which produces the ultrasound field. The authors demonstrated extremely fast changes (< 0.1 s) in the motor speed and reproducible "on/off" activations that were faster than those by using other reported methods for stopping the propulsion of microjets, as shown in **Figure 15B** [45].

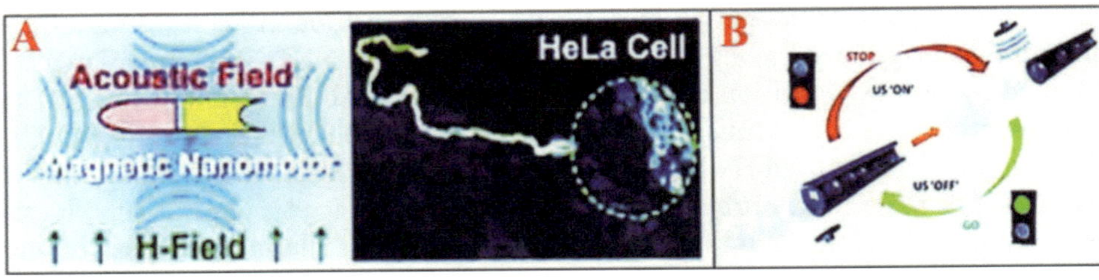

Figure 15.
(A) Scheme representing controlling of acoustically propelled nanowire toward a HeLa cell. (B) Scheme representing ultrasound-modulated bubble propulsion of chemically powered microtubes. Copyright 2013, ACS Publications. Copyright 2014, ACS Publications.

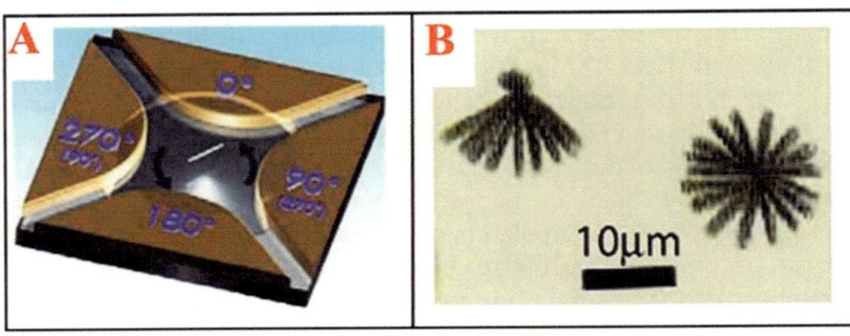

Figure 16.
Rotation of micro/nanomotors by applying AC voltages to multiple electrodes: (A) schematic diagram of experimental setup of quadruple electrodes and (B) images of one end fixed (left) and free (right) rotating Au nanowires, respectively. Copyright 2005, APS Publications.

4.3 Electric control

Metallic micro-/nanomotors can perform controllable rotation resulting from rotational torque in an electric field provided by applying AC voltages to multiple electrodes, as shown in **Figure 16A** [46]. In addition, by applying AC electric fields to strategically designed microelectrodes, the propulsion of metallic micro-/nanomotors could be tuned by dielectrophoretic force. They could be driven to chain, accelerate and align in certain directions, as well as to disperse, concentrate and assemble into complex scaffolds, as shown in **Figure 16B**.

4.4 Light control

Using the light to guide the movement of micro-/nanomotors was also reported. For example, Solovev and collaborators studied the control of microjets by using a white light source, as shown in **Figure 17A** [47]. This process was mediated through the illumination of the fuel solution above Pt-patterned Si surfaces, which generates a local decrease of the surfactant and H_2O_2 concentration. Although white light could be applied to switch off the motion of the microjets, shorter wavelengths were attributed to suppress the production of microbubbles faster than longer wavelengths. The phenomenon can be reversible, and thus a nonactive microjet is triggered by dimming the light source. Nevertheless, the "on/off" process is not immediate since it demands a few seconds to completely terminate or to reach a constantly maximum speed.

He and his colleagues recently reported the near infrared (NIR) light-switching "on/off" propulsion of Au nanoshell-functionalized polymer multilayer rockets, as

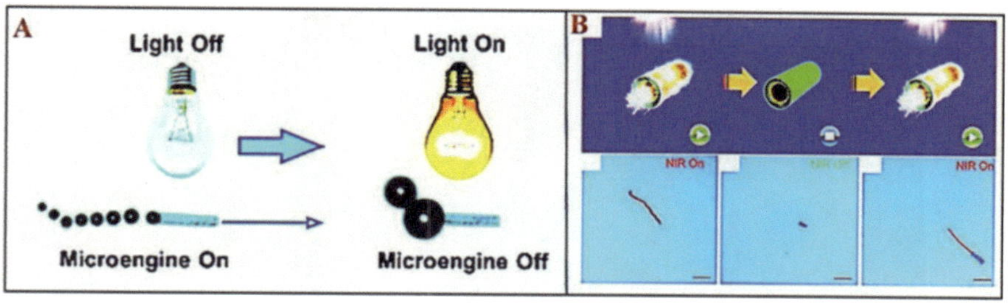

Figure 17.
(A) Microengine's propulsion controlled by light. (B) NIR light-switchable motion of NIR propelled polymer multilayer rockets. Copyright 2011, Wiley Online Library. Copyright 2010, Wiley Online Library.

shown in **Figure 17B** [48]. NIR light irradiation could rapidly generate a thermal gradient that enables reversible movement of the Au nanoshell-functionalized polymer multilayer rockets. The rockets exhibit "on/off/on" cycles in response to an adjustable NIR irradiation, along with "go/stop/go" motions. The directional movements of the rockets were terminated as the NIR irradiation is "off" and were resumed upon switching on the NIR light. Accounting for the straight propulsion behavior of NIR-propelled polymer multilayer rockets, the "on/off" NIR switching realizes the precisely predefined route of the polymer multilayer rockets. In addition, the rockets were sustained with negligible damage under 30 times of NIR irradiation and were highly durable.

4.5 Thermal control

Thermal control of the propulsion of artificial micro-/nanomotors has proved to be applicable for both micro-/nanowires and microtubes. The speed of Pt-Au nanowires was substantially increased upon exposure to elevated temperatures. Similar phenomenon was discovered for bubble-propelled microtubes, which has been used to compensate the effect of decreasing the fuel level.

The temperature of the solution could be controlled by two Peltier elements in connection with a direct current (DC) power supply placed below the sample containing microjets. By heating up the system to a physiological temperature, microjets increase their efficiency and are propelled at extremely low concentrations of H_2O_2, as shown in **Figure 18** [49, 50]. In addition, soft micromotors consisted of flexible thermo-responsive polymeric microjets could reversibly fold and unfold in an accurate manner resulted from the temperature change of the solution in which they are dispersed, thereby allowing them to rapidly initiate and terminate multiple times in response to the radius of curvature accordingly. The employment of stimuli-responsive materials would be ideal for the future designs of smart micro-/nanomotors.

4.6 Chemical control

The propulsion of self-propelled micro-/nanomotors could be adjusted by tuning the fuel level or chemical stimuli. In the presence of fuel concentration gradient, micro-/nanomotors can drive themselves along the gradient toward a region with a higher fuel concentration. As such, monitoring the fuel concentration and distribution can be used to direct and modulate the propulsion of micro-/nanomotors. In addition to fuel concentration, the propulsion of micro-/nanomotors is influenced by the presence of certain other chemicals. For instance, the movement of Au/Pt nanomotors was reported to be extremely accelerated upon the addition of Ag ion because of the under-potential deposition of Ag on the nanomotors, which

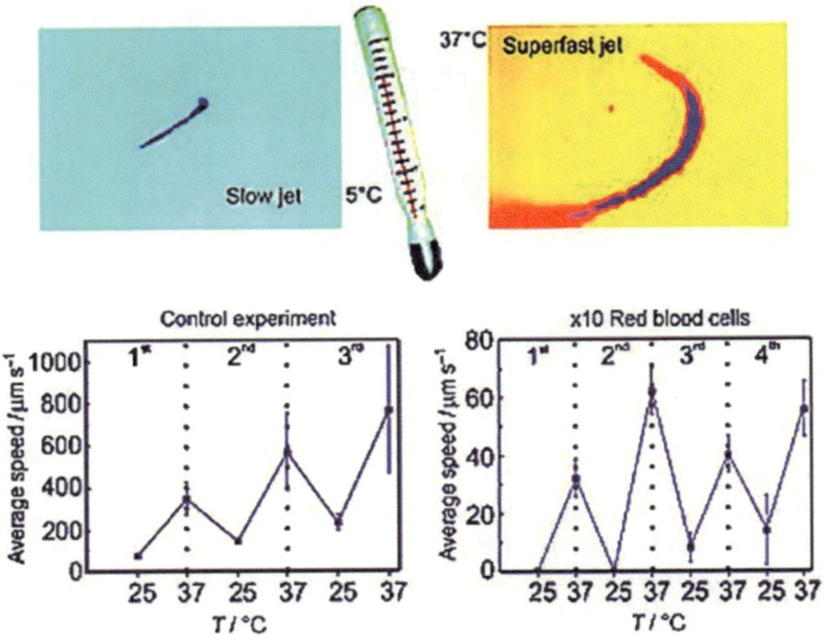

Figure 18.
Micro/nanojet's propulsion controlled by temperature. Copyright 2011, ACS Publications. Copyright 2013, RSC Publications.

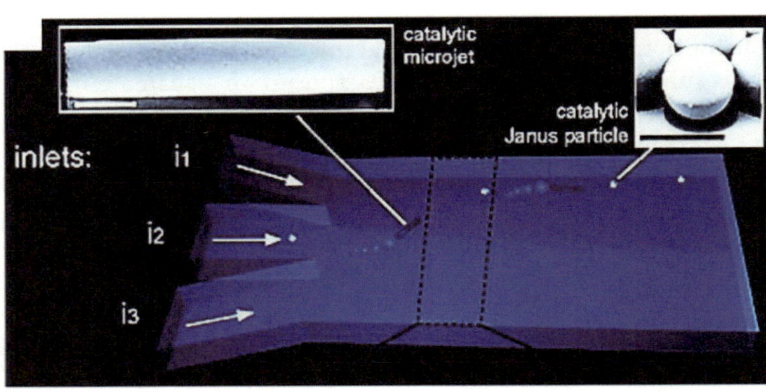

Figure 19.
Micro/nanomotor's motion controlled by chemical gradient in microfluidic channel. Copyright 2013, Wiley Online Library.

introduces differences in surface and catalytic properties. Hydrazine (N_2H_4) is another chemical stimulus observed to be effective to accelerate the propulsion of Au/Pt-CNT nanomotors. For bubble-propelled micro-/nanomotors, surfactants are significantly critical to the mobility of motors, resulting from they can stimulate bubble production and detachment.

Solovev et al. reported that the production of large microbubbles from small ensembles of microjets generated a chemophoretic attraction force and a capillary force that pulled other microjets into the swarm. A more complicated experiment was demonstrated by Baraban et al., who reported a controllable manner to study the chemotactic behavior of Janus motors and tubular microjets in microfluidic channels. Both types of motors move toward the gradient of the fuel without the influence of capillary forces, as shown in **Figure 19** [51].

5. Applications

There are many current and potential applications, resulting from the great advances in cargo-towing force, propulsion control, and lifetime of synthetic micro-/nanomotors. The wide range of potential applications of micro-/nanomotors covering different fields requires specific functionalization strategies in each kind of application. Herein, the functionalization of micro-/nanomotors for four main categories of applications is reported as follows: cargo delivery, environmental remediation, chemical sensors, and biomedical applications.

5.1 Cargo delivery

Cargo delivery is one of the most important envisioned applications of micro-/nanomotors. Relying on the properties of cargoes, tailored methods are desired for their corresponding delivery. A general process of how to synthesize micro-/nanomotors and employ them to deliver cargo molecules has been shown in **Figure 20** [4].

For the cargo delivery by micro-/nanomotors, the cargo could simply be connected to the motors by magnetic attraction. The delivery of drug-loaded magnetic poly(D,L-lactic-co-glycolic acid) (PLGA) microparticles has been studied by both chemically propelled, as shown in **Figure 21A** [52–54], and magnetically driven micro-/nanomotors. For charged cargoes, electrostatic interaction between cargoes and micro-/nanomotors could be applied for the pick-up process. A common scheme introducing charged portions into micro-/nanomotors is to incorporate a negatively charged polymer part. Sen et al. reported that a PPy part was incorporated to a nanowire via electropolymerization, which could be attached to oppositely charged polystyrene amidine cargo via electrostatic interaction, as shown in **Figure 21A**. A photo-chemically triggered cargo unloading manner was proposed for cargoes loaded nanowires via electrostatic interaction. An additional Ag portion

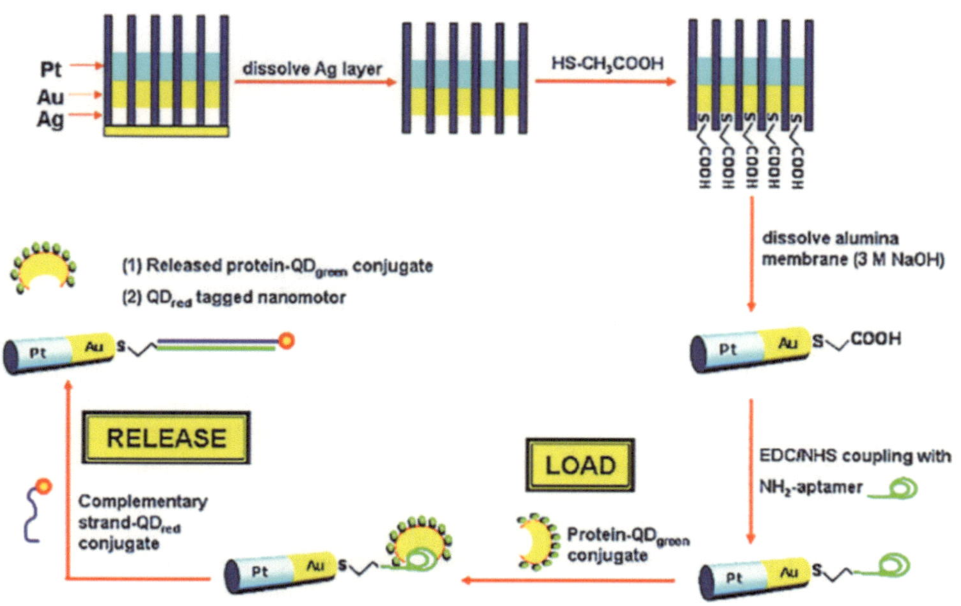

Figure 20.
Electrodeposited aptamer-functionalized micro/nanomotors for selective loading, deliver, and unloading of a protein cargo. Copyright 2009, ACS Publications.

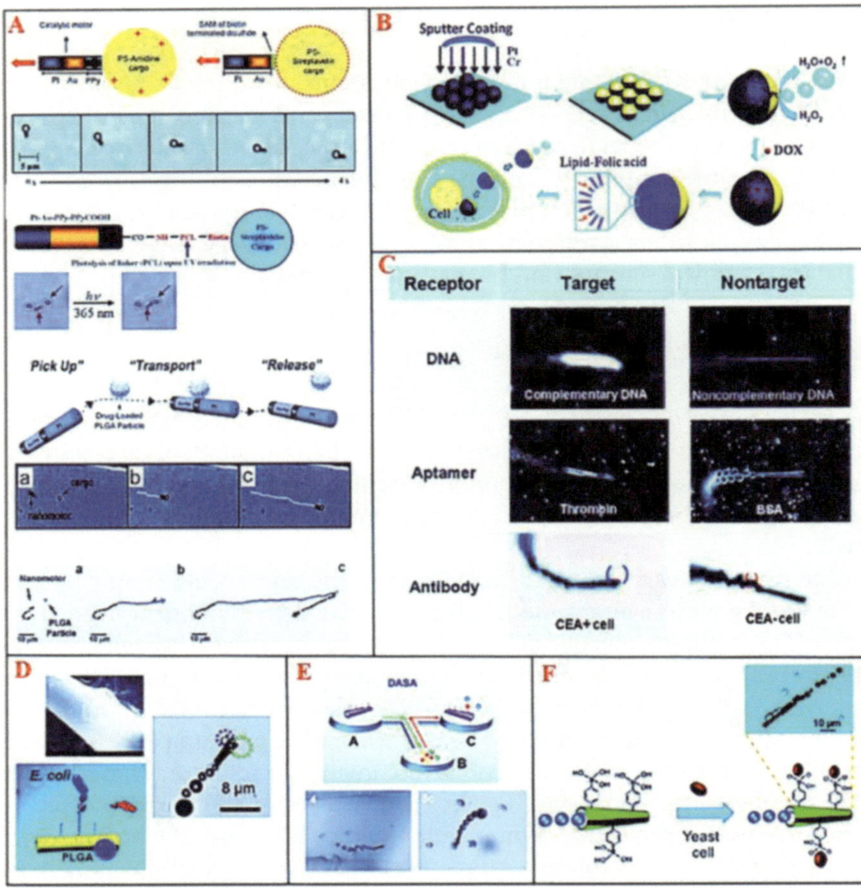

Figure 21.
(A) Examples of the delivery of cargo using solid micro/nanorods: (a) cargo pick-up, (b) cargo delivery, and (c) cargo release, respectively. (B) Synthesis procedures for Janus spherical micro/nanomotors, drug picking-up, lipid bilayer functionalization, and drug unloading (DOX = doxorubicin hydrochloride). (C) Selective binding and delivery of biological analytes and cells by functionalized microjets. (D) Direct optical visualization of pick-up, transportation, and delivery of E. coli bacteria and polymeric drug-carrier spheres. (E) In-chip immunoassays for in situ picking up and delivery of target proteins. (F) Selective recognition of monosaccharides for loading and unloading of yeast cells. Copyright 2008, ACS Publications. Copyright 2010, Wiley Online Library. Copyright 2010, Wiley Online Library. Copyright 2014, Wiley Online Library. Copyright 2011, RSC Publications. Copyright 2012, ACS Publications. Copyright 2013, RSC Publications. Copyright 2012, ACS Publications.

in a nanowire will be dissolved rapidly in the presence of H_2O_2, chloride ions (Cl^-), and ultraviolet (UV) light, resulting in releasing of the cargo.

Garcia-Gradilla and colleagues demonstrated that the incorporation of a negatively charged polypyrrole polystyrene sulfonate (PPyPSS) portion with an ultrasound-propelled nanowire could be served as a pH-sensitive carrier for positively charged drugs via electrostatic interaction. The unloading of the drugs was theoretically realized by a protonated PPyPSS portion in an acidic environment. The same group also reported drug-loaded nanowires depending on a nanoporous Au portion with a large surface area. Such a nanoporous device is synthesized by dealloying the Ag portion of an Au/Ag alloy grown by the coelectrodeposition of Au and Ag. The picking up of the drug doxorubicin in the nanopores via electrostatic interactions with the polymeric coating of the nanowire motors and NIR light-stimulated drop off were reported as well.

Xuan et al. reported the fabrication of self-propelled Janus nanomotors in a diameter of 75 nm. The steps are illustrated in **Figure 21B** [55]. Spherical mesoporous SiO_2 nanoparticles (MSNs) were fabricated by using a base-catalyzed sol–gel method. Afterwards, MSNs were dispersed on an Si substrate to form a monolayer and subsequently deposited with evaporated Cr (chromium) and Pt. A brief sonication resulted in the Janus MSNs becoming spread in a solution of the anticancer drug doxorubicin hydrochloride to pick up the cargo, and subsequently mixing with 1-mg mL^{-1} egg phosphatidylcholine containing 1% folic acid resulted in encapsulation.

The Wang's group reported a great capability of microtubular motors for the selective loading, delivery, and isolation of distinct target analytes of biological relevance. They bio-functionalized the outer walls of microtubes with antibodies, aptamers, lectin receptors, and ss-DNA, and thus the isolation modes of cancer cells, bacteria, proteins, and nucleic acids could be demonstrated in **Figure 21C** [56]. The authors made use of rolled-up microtubes with Au layers for thiol modification and template-assisted microjets with polymer walls.

Campuzano and his colleagues functionalized Au/Ni/PANI/Pt micromotors with concanavalin A (ConA) to enable selective loading of pathogenic bacteria from fuel-enhanced real samples, as shown in **Figure 21D** [57]. The loading, transport, and unloading events were observed by optical microscopy. The delivery of pathogenic bacteria by magnetic polymeric drug carriers could represent the basis of an attractive propulsion-based theranostics scheme. In related work, the same group later functionalized artificial catalytic micromotors with antibodies to enable in-chip immunoassays, as shown in **Figure 21E** [58]. Kuralay et al. fabricated poly(3-aminophenylboronic acid) (PAPBA)/Ni/Pt tubular micromotors that selectively recognize monosaccharides, which were capable of loading and drop off of yeast cells, as shown in **Figure 21F** [59].

5.2 Environmental remediation

One of the primary environmental applications of micro-/nanomotors is to adsorb the pollutants in water. Remediation agents could be incorporated with micro-/nanomotors as the outer surface to contribute to the purification process during propulsion. Soler et al. studied the application of microtube motors decorated with a Fe outer surface to degrade organic contaminants in water via the Fenton oxidation, as shown in **Figure 22A** [60]. Wang et al. reported the application of PEDOT/Pt microtubular motors to promote the degradation of chemical threats, as shown in **Figure 22B** [61]. In brief, the oxidation of an organophosphate nerve agent by H_2O_2 was enhanced in the presence of the self-propelled micromotors, which resulted in an efficient mixing of the treated aqueous solution without the aid of external mechanical stirrers.

Surface modifications of micro-/nanomotors with a hydrophobic layer could also activate them to load oil droplets. Guix et al. demonstrated that the functionalized Au/Ni/PEDOT/Pt microtubes with a self-assembled monolayer (SAM) of alkanethiols on the Au outer surface could pick up and deliver oil droplets resulting from strong interactions between them, as shown in **Figure 22C** [62]. The same group also incorporated the rough external Au layer of microengines with long chains of self-assembled monolayers to create a super-hydrophobic absorbent layer for oil loading, as shown in **Figure 22D** [62].

5.3 Chemical sensors

The application of micro-/nanomotors as chemical sensors is based on the case that the propulsion speed of micro-/nanomotors can be converted into an analytically useful signal. The interaction of certain compounds in the sample with the

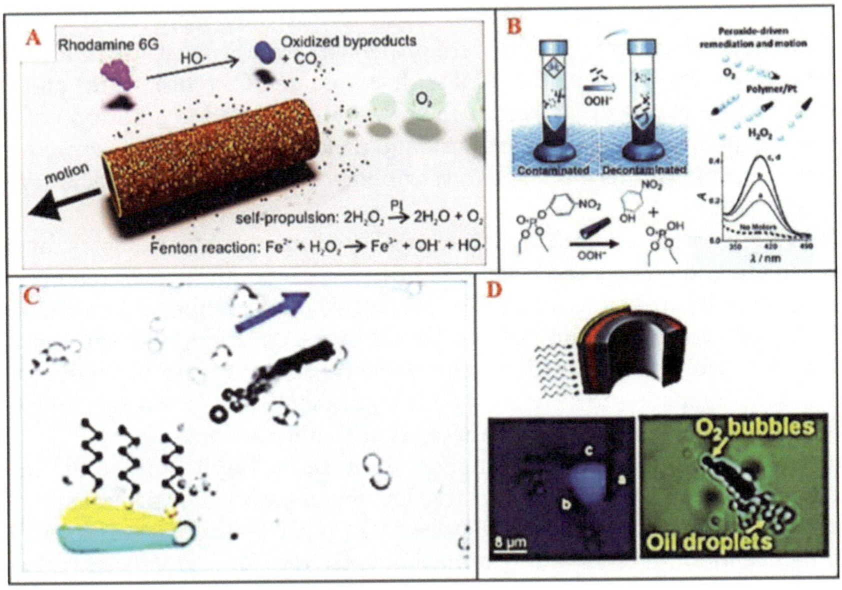

Figure 22.
(A) Organic pollutants degraded by multifunctional Fe/Pt micromotors in H_2O_2 solutions. (B) Accelerated oxidation of organophospate nerve agents by using micromotors as active mixers. (C) Loading of oil droplets by alkanethiol-modified microtubes. (D) Picking up and loading of oil, enabled by a superhydrophobic-modified outer layer. Copyright 2013, ACS Publications. Copyright 2013, Wiley Online Library. Copyright 2012, ACS Publications.

catalytic sites of micro-/nanomotors leads to the alteration of their propulsion speed and is related to the concentration of an analyte in solution, as shown in **Figure 23** [63–66]. Although the field is still in its infancy, micro-/nanomotors as chemical sensors could have a number of advantages over conventional either optical or electrochemical sensors, such as sensitivity, selectivity, immunity to electrical interferences, operation in a wireless manner, and only requiring a minute amount of sample. In the past few years, some research groups demonstrated the ability of micro-/nanomotors to detect inorganic electrolytes present in blood, heavy metals, organic compounds such as dimethyl sulfoxide (**Figure 23A**) [63], uric acid, blood proteins such as bovine serum albumin (BSA), glucose oxidase enzymes and g-globulin, amino acids containing thiol groups, for instance, methionine, cysteine, and serine (**Figure 23D**) [65], peptides such as glutathione, and DNA (**Figure 23A**).

García and his colleagues demonstrated the first antibody-loaded tubular microengines, which were developed to load and deliver target molecules between different microfluidic chambers. Catalytic polymer/Ni/Pt microengines were biofunctionalized with antibodies targeting Immunoglobulin G (IgG) protein molecule, as a model protein, in order to realize a micromotor-based immunoassay providing "on-the-fly" loading and isolation/sorting capabilities, as shown in **Figure 23B** [57]. The immunocomplex could be simply observed by optical microscope through using an antigen/antibody labeled with a polymeric sphere tracer. This innovative work is highly selective and excludes time-consuming washing steps, accelerating and simplifying the general immunoassay procedures. Taking advantage of these features, Yu et al. employed antibody-loaded AuNP/PANI/Pt micromotors to exhibit rapid "on-the-fly" sandwich immunocomplexes targeting carcinoembryonic antigen. The operation takes 5 minutes with a measuring threshold of 1–1000 ng mL^{-1}. In addition, labeling the loaded proteins with microscopic particles demonstrating different sizes and shapes facilitates the multiplexed analysis of proteins, as proved by Vilela and his colleagues (**Figure 23C**) [64]. In

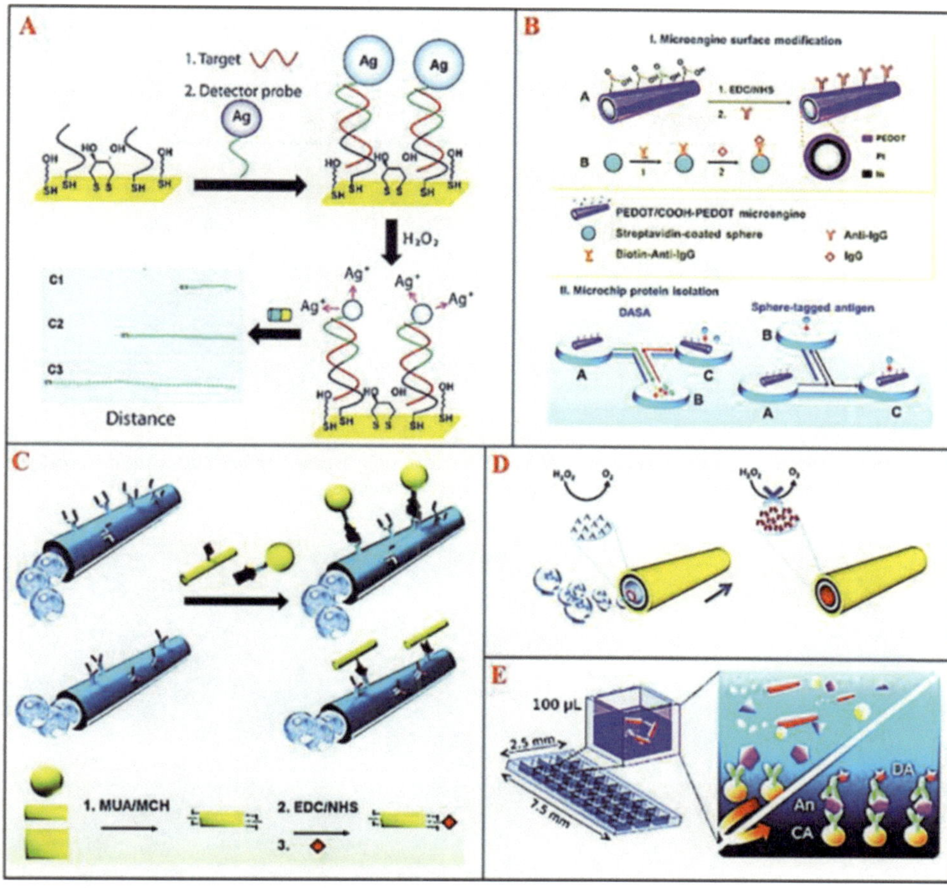

Figure 23.
(A) Detection of Ag-tagged nucleic acid, which alters the propulsion of the micro-/nanomotors. (B) Antibody-decorated micromotor for protein detection. (C) Micromotor-based multiplexed immunoassay via different microscopic tracers. (D) Effect of the concentration of DMSO (dimethyl sulfoxide), cysteine, and serine on the swimming speeds of the microtubular motors. (E) Microarray immunoassay assisted by microengines. Copyright 2013, ACS Publications. Copyright 2012, ACS Publications. Copyright 2014, RSC Publications. Copyright 2013, RSC Publications. Copyright 2014, Wiley Online Library.

addition, the aforementioned vortex effect generated by the propulsion of micromotors can boost the mass transfer of the target toward the functionalized microjet surface (i.e. "on-the-fly") as well as assist the mass transfer of the target molecule within the matrix of a sample solution toward a sensing surface, where the bio-receptor is situated and the target is expected to be selectively attached, as shown in **Figure 23E** [66].

5.4 Biomedical applications

The pioneering micro-/nanomotors offer considerable promise for biomedical applications, as shown in **Figure 24** [67]. Herein, we outline some important advances of micro-/nanomotors in the biomedical field.

Micro-/nanojets have proven to be capable of drilling into biomaterials and soft tissues. Rolled-up thin nanomembranes can asymmetrically result in sharp edges being engineered. Nanojets were self-propelled and externally directed toward immobilized cancer cells and embedded in their interior, as shown in **Figure 25A** [68]. However, the toxicity of the H_2O_2 fuel used for the movement leads to the cells undergoing apoptosis after short periods. Therefore, other environmentally

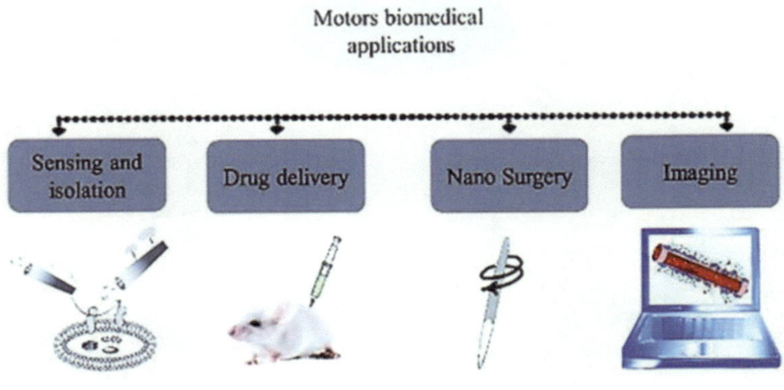

Figure 24.
Categories of biomedical applications of micro/nanomotors, including sensing and isolation, drug delivery, nanosurgery, biomedical imaging. Copyright 2014, RSC Publications.

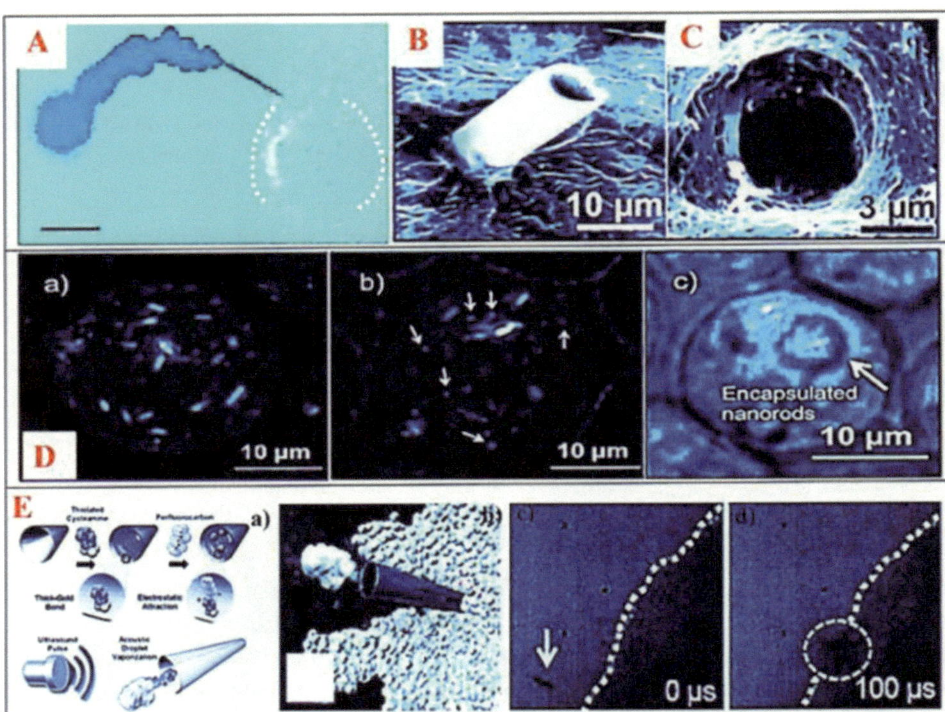

Figure 25.
(A) A catalytic nanojet drilling into an immobilized cancer cell. (B) SEM image of a magnetic microtube drilling into pig liver tissue. (C) Hole after removing the microtube. (D) Nanorods inside living HeLa cells: (a) multiple nanorods inside a HeLa cell, (b) subcellular structures (a few indicated by arrows) interacting with active acoustic motors inside a HeLa cell, and (c) two vesicular structures in a HeLa cell containing many active but crowded acoustic motors, respectively, can be seen. (E) Preparation and propulsion of perfluorocarbon-loaded microbullets (a), Computer-created diagram (b), and images of a microbullet penetrating a tissue after an ultrasound pulse (c and d), respectively. Copyright 2012, ACS Publications. Copyright 2013, RSC Publications. Copyright 2014, Wiley Online Library. Copyright 2012, Wiley Online Library.

friendly sources of movement are urgently required to replace H_2O_2. Magnetic trigger can be used as an alternative to the motion. Conical microtubes powered by an external rotating magnetic field were placed onto liver tissue and drilled into it for long periods, as shown in **Figure 25B** [69]. After releasing the microtube from the biomaterial, a micrometer-sized hole was observed by SEM, as shown in

Figure 25C [69]. Ultrasonic waves can also generate the propulsion of micro-/nanomotors, as shown in **Figure 25D** and **E**) [70, 71]. Bio-functionalized nanowires propelled by ultrasound have been applied for bio-sensing, and the first results of magnetic steering toward cells have been studied. Mallouk et al. demonstrated the internalization of a nanowire-based motor inside living cells, as shown in **Figure 25D**. Developing biocompatible materials and fuels for artificial micro-/nanomotors or the use of noninvasive external triggered motors may pave the way for biomedical applications of micro-/nanomotors in the near future [72–80].

6. Conclusions

In this chapter, the previously reported Pt-based micro-/nanomotors are presented. In detail, the propulsion mechanisms, fabrication methods, propulsion controlling methods, and applications of these synthetic micro-/nanomotors developed in the past years are summarized accordingly. Despite the rapid and significant advances in micro-/nanomotors, challenges such as specifically practical applications and smart controlling still remain to be resolved. In addition, advanced structure design and fabrication methods are demanded.

Acknowledgements

This work was supported by Ministry of Education, Singapore, under "MOE 2011-T2-2-156, ARC 18/12" program.

Conflicts of interest

There are no conflicts of interest to declare.

Author details

Liangxing Hu[1]*, Nan Wang[2] and Kai Tao[3]

1 Micromachines Lab, School of Mechanical and Aerospace Engineering, Nanyang Technological University, Singapore

2 Department of Science and Math, Singapore University of Technology and Design, Singapore

3 Ministry of Education Key Laboratory of Micro and Nano Systems for Aerospace, School of Mechanical Engineering, Northwestern Polytechnical University, Xi'an, China

*Address all correspondence to: lhu001@e.ntu.edu.sg

References

[1] Stock D, Namba K, Lee LK. Nanorotors and self-assembling macromolecular machines: The torque ring of the bacterial flagellar motor. Current Opinion in Biotechnology. 2012;**23**:545

[2] Schmidt JJ, Montemagno CD. Hybrid protein-polymer biomimetic membranes. Annual Review of Materials Research. 2004;**34**:315

[3] Kinosita K, Yasuda R, Noji H, Adachi K. A rotary molecular motor that can work at near 100% efficiency. Philosophical Transactions of the Royal Society B: Biological Sciences. 2000;**355**:473

[4] Wang J. Can man-made nanomachines compete with nature biomotors? ACS Nano. 2009;**3**:4

[5] Wang W, Castro LA, Hoyos M, Mallouk TE. Autonomous motion of metallic microrods propelled by ultrasound. ACS Nano. 2012;**6**:6122

[6] Wang Y, Hernandez RM, Bartlett DJ Jr, Bingham JM, Kline TR, Sen A, et al. Bipolar electrochemical mechanism for the propulsion of catalytic nanomotors in hydrogen peroxide solutions. Langmuir. 2006;**22**:10451

[7] Wang J et al. Multi-fuel driven Janus micromotors. Small. 2013;**9**:467

[8] Mirkovic T, Zacharia NS, Scholes GD, Ozin GA. Nanolocomotion-catalytic Nanomotors and nanorotors. Small. 2010;**6**:159

[9] Patra D, Sengupta S, Duan W, Zhang H, Pavlick R, Sen A. Intelligent, self-powered, drug delivery systems. Nanoscale. 2013;**5**:1273

[10] Whitesides GM et al. Autonomous movement and self-sssembly. Angewandte Chemie, International Edition. 2002;**41**:652

[11] Gibbs JG, Zhao YP. Design and characterization of rotational multicomponent catalytic nanomotors. Small. 2009;**5**:2304

[12] Gibbs J, Zhao Y. Catalytic nanomotors: fabrication, mechanism, and applications. Frontiers of Materials Science. 2011;**5**:25

[13] Paxton WF, Kistler KC, Olmeda CC, Mallouk TE, Sen A. Catalytic nanomotors: Autonomous movement of striped nanorods. Journal of the American Chemical Society. 2004;**126**: 13424

[14] Jiang HR, Yoshinaga N, Sano M. Active motion of a Janus particle by self-thermophoresis in a defocused laser beam. Physical Review Letters. 2010; **105**:268302

[15] Baraban L, Streubel R, Makarov D, Han L, Karnaushenko D, Schmidt OG, et al. Catalytic Janus motors on microfluidic chip: deterministic motion for targeted cargo delivery. ACS Nano. 2012;**7**:1360

[16] Mei YF, Solovev AA, Sanchez S, Schmidt OG. Rolled-up nanotech on polymers: From basic perception to self-propelled catalytic microengines. Chemical Society Reviews. 2011;**40**: 2109

[17] Gibbs JG, Zhao YP. Autonomously motile catalytic nanomotors by bubble propulsion. Applied Physics Letters. 2009;**94**(163103):1

[18] Wang H, Pumera M. Fabrication of micro/nanoscalemotors. Chemical Reviews. 2015;**115**:8704

[19] Demirok UK, Laocharoensuk R, Manesh KM, Wang J. Ultrafast catalytic alloy nanomotors. Angewandte Chemie, International Edition. 2008;**47**:9349

[20] Laocharoensuk R, Burdick J, Wang J. Carbon-nanotuble-induced acceleration of catalytic nanomotors. ACS Nano. 2008;**2**:1069

[21] Gao W, Sattayasamitsathit S, Orozco J, Wang J. Highly efficient catalytic microengines: template electrosynthesis of polyaniline/platinum microtubes. Journal of the American Chemical Society. 2011;**133**:11862

[22] Mirkovic T, Foo ML, Arsenault AC, Fournier-Bidoz S, Zacharia NS, Ozin GA. Hinged nanorods made using a chemical approach to flexible nanostructures. Nature Nanotechnology. 2007;**2**:565

[23] Zhao G, Ambrosi A, Pumera M. Self-propelled nanojets via template electrodeposition. Nanoscale. 2013;**5**:1319

[24] Gao W, Sattayasamitsathit S, Manesh KM, Weihs D, Wang J. Magnetically powered flexible metal nanowire motors. Journal of the American Chemical Society. 2010;**132**:14403

[25] Wheat PM, Marine NA, Moran JL, Posner JD. Rapid fabrication of bimetallic spherical motors. Langmuir. 2010;**26**:13052

[26] Valadares LF, Tao YG, Zacharia NS, Kitaev V, Galembeck F, Kapral R, et al. Catalytic nanomotors: self-propelled sphere dimers. Small. 2010;**6**:565

[27] Gibbs JG, Fragnito NA, Zhao Y. Asymmetric Pt/Au coated catalytic micromotors fabricated by dynamic shadowing growth. Applied Physics Letters. 2010;**97**:253107

[28] Lee TC, Alarcon-Correa M, Miksch C, Hahn K, Gibbs JG, Fischer P. Self-propelling nanomotors in the presence of strong Brownian forces. Nano Letters. 2014;**14**:2407

[29] He Y, Wu J, Zhao Y. Designing catalytic nanomotors by dynamic shadowing growth. Nano Letters. 2007; 7:1369

[30] Huang W, Manjare M, Zhao Y. Catalytic nanoshell micromotors. Journal of Physical Chemistry C. 2013; **117**:21590

[31] Solovev AA, Mei Y, Bermudez Urena E, Huang G, Schmidt OG. Catalytic microtubular jet engines self-propelled by accumulated gas bubbles. Small. 2009;**5**:1688

[32] Solovev AA, Sanchez S, Pumera M, Mei YF, Schmidt OG. Magnetic control of tubular catalytic microbots for the transport, assembly, and delivery of micro-objects. Advanced Functional Materials. 2010;**20**:2430

[33] Yao K, Manjare M, Barrett CA, Yang B, Salguero TT, Zhao Y. Nanostructured scrolls from graphene oxide for microjet engines. Journal of Physical Chemistry Letters. 2012;**3**:2204

[34] Magdanz V, Stoychev G, Ionov L, Sanchez S, Schmidt OG. Stimuli-responsive microjets with reconfigurable shape. Angewandte Chemie, International Edition. 2014;**53**: 2673

[35] Li J, Zhang J, Gao W, Huang G, Di Z, Liu R, et al. Dry-released nanotubes and nanoengines by particle-assisted rolling. Advanced Materials. 2013;**25**:3715

[36] Wu Y, Wu Z, Lin X, He Q, Li J. Autonomous movement of controllable assembled Janus capsule motors. ACS Nano. 2012;**6**:10910

[37] Wilson DA, Nolte RJM, van Hest JCM. Autonomous movement of platinum-loaded stomatocytes. Nature Chemistry. 2012;**4**:268

[38] Garcia-Gradilla V, Orozco J, Sattayasamitsathit S, Soto F, Kuralay F, Pourazary A, et al. Functionalized

ultrasound-propelled magnetically guided nanomotors: toward practical biomedical applications. ACS Nano. 2013;7:9232

[39] Wu Y, Lin X, Wu Z, Möhwald H, He Q. Self-propelled polymer multilayer Janus capsules for effective drug delivery and light-triggered release. ACS Applied Materials & Interfaces. 2014;**6**: 10476

[40] Kline TR, Paxton WF, Mallouk TE, Sen A. Catalytic nanomotors: Remote-controlled autonomous movement of striped metallic nanorods. Angewandte Chemie, International Edition. 2005;**44**: 744

[41] Zhao G, Pumera M. Magnetotactic artificial self-propelled nanojets. Langmuir. 2013;**29**:7411

[42] Baraban L, Makarov D, Streubel R, Mönch I, Grimm D, Sanchez S, et al. Catalytic Janus motors on microfluidic chip: deterministic motion for targeted cargo delivery. ACS Nano. 2012;**6**:3383

[43] Khalil ISM, Magdanz V, Sanchez S, Schmidt OG, Misra S. Three-dimensional closed-loop control of self-propelled microjets. Applied Physics Letters. 2013;**103**:172404

[44] Ahmed S, Wang W, Mair LO, Fraleigh RD, Li S, Castro LA, et al. Steering acoustically propelled nanowire motors toward cells in a biologically compatible environment using magnetic fields. Langmuir. 2013;**29**:16113

[45] Xu T, Soto F, Gao W, Garcia-Gradilla V, Li J, Zhang X, et al. Ultrasound-modulated bubble propulsion of chemically powered microengines. Journal of the American Chemical Society. 2014;**136**:8552

[46] Fan DL, Zhu FQ, Cammarata RC, Chien CL. Controllable high-speed rotation of nanowires. Physical Review Letters. 2005;**94**:247208

[47] Solovev AA, Smith EJ, Bufon CCB, Sanchez S, Schmidt OG. Light-controlled propulsion of catalytic microengines. Angewandte Chemie, International Edition. 2011;**50**:10875; Angewandte Chemie. 2011;**123**:11067

[48] Hong Y, Diaz M, Córdova-Figueroa UM, Sen A. Light-driven titanium dioxide based reversible microfireworks and micro-motor/micropump systems. Advanced Functional Materials. 2010; **20**:1568

[49] Sanchez S, Ananth AN, Fomin VM, Viehrig M, Schmidt OG. Superfast motion of catalytic microjet engines at physiological temperature. Journal of the American Chemical Society. 2011; **133**:14860

[50] Soler L, Martinez-Cisneros C, Swiersy A, Sanchez S, Schmidt OG. Thermal activation of catalytic microjets in blood samples using microfluidic chips. Lab on a Chip. 2013; **13**:4299

[51] Baraban L, Harazim SM, Sanchez S, Schmidt OG. Chemotactic behavior of catalytic motors in microfluidic channels. Angewandte Chemie, International Edition. 2013;**52**:5552; Angewandte Chemie. 2013;**125**:5662

[52] Sundararajan S, Lammert PE, Zudans AW, Crespi VH, Sen A. Catalytic motors for transport of colloidal cargo. Nano Letters. 2008;**8**: 1271

[53] Sundararajan S, Sengupta S, Ibele ME, Sen A. Drop-off of colloidal cargo transported by catalytic Pt-Au nanomotors via photochemical stimuli. Small. 2010;**6**:1479

[54] Kagan D, Laocharoensuk R, Zimmerman M, Clawson C, Balasubramanian S, Kong D, et al. Rapid delivery of drug carriers propelled and navigated by catalytic nanoshuttles. Small. 2010;**6**:2741

[55] Xuan M, Shao J, Lin X, Dai L, He Q. Self-propelled Janus mesoporous silica nanomotors with sub-100 nm diameters for drug encapsulation and delivery. ChemPhysChem. 2014;**15**:2255

[56] Campuzano S, Kagan D, Orozco J, Wang J. Motion-driven sensing and biosensing using electrochemically propelled nanomotors. Analyst. 2011; **136**:4621

[57] Campuzano S, Orozco J, Kagan D, Guix M, Gao W, Sattayasamitsathit S, et al. Bacterial isolation by lectin-modified microengines. Nano Letters. 2012;**12**:396

[58] García M, Orozco J, Guix M, Gao W, Sattayasamitsathit S, Escarpa A, et al. Micromotor-based lab-on-chip immunoassays. Nanoscale. 2013;**5**:1325

[59] Kuralay F, Sattayasamitsathit S, Gao W, Uygun A, Katzenberg A, Wang J. Self-propelled carbohydrate-sensitive microtransporters with built-in boronic acid recognition for isolating sugars and cells. Journal of the American Chemical Society. 2012;**134**:15217

[60] Soler L, Magdanz V, Fomin VM, Sanchez S, Schmidt OG. Self-propelled micromotors for cleaning polluted water. ACS Nano. 2013;**7**:9611

[61] Orozco J, Cheng GZ, Vilela D, Sattayasamitsathit S, Vazquez-Duhalt R, Valdes-Ramirez G, et al. Micromotor-based high-yielding fast oxidative detoxification of chemical threats. Angewandte Chemie, International Edition. 2013;**52**:13276; Angewandte Chemie. 2013;**125**:13518

[62] Guix M, Orozco J, Garcia M, Gao W, Sattayasamitsathit S, Merkoci A, et al. Superhydrophobic alkanethiol-coated microsubmarines for effective removal of oil. ACS Nano. 2012;**6**:4445

[63] Campbell AI, Ebbens SJ. Gravitaxis in spherical Janus swimming devices. Langmuir. 2013;**29**:14066

[64] Vilela D, Orozco J, Cheng G, Sattayasamitsathit S, Galarnyk M, Kan C, et al. Multiplexed immunoassay based on micromotors and microscale tags. Lab on a Chip. 2014;**14**:3505

[65] Zhao GJ, Sanchez S, Schmidt OG, Pumera M. Poisoning of bubble propelled catalytic micromotors: the chemical environment matters. Nanoscale. 2013;**5**:2909

[66] Morales-Narváez E, Guix M, Medina-Sánchez M, Mayorga-Martinez CC, Merkoçi A. Micromotor enhanced microarray technology for protein detection. Small. 2014;**10**:2542

[67] Loai KEA, Abdelmohsen F, Peng YFT, Wilson DA. Micro- and nano-motors for biomedical applications. Journal of Materials Chemistry B. 2014;**2**:2395

[68] Solovev AA, Xi W, Gracias DH, Harazim SM, Deneke C, Sanchez S, et al. Self-propelled nanotools. ACS Nano. 2012;**6**:1751

[69] Xi W, Solovev AA, Ananth AN, Gracias DH, Sanchez S, Schmidt OG. Rolled-up magnetic microdrillers: towards remotely controlled minimally invasive surgery. Nanoscale. 2013;**5**: 1294

[70] Wang W, Li SX, Mair L, Ahmed S, Huang TJ, Mallouk TE. Acoustic propulsion of nanorod motors inside living cells. Angewandte Chemie, International Edition. 2014;**53**:3201; Angewandte Chemie. 2014;**126**:3265

[71] Kagan D, Benchimol MJ, Claussen JC, Chuluun-Erdene E, Esener S, Wang J. Acoustic droplet vaporization and propulsion of perfluorocarbon-loaded microbullets for targeted tissue penetration and deformation. Angewandte Chemie, International Edition. 2012;**51**:7519; Angewandte Chemie. 2012;**124**:7637

[72] Hu LX, Miao JM, Grüber G. Disk-like nanojets with steerable trajectory using platinum nozzle nanoengines. RSC Advances. 2016;**6**:3399

[73] Hu LX, Miao JM, Grüber G. Temperature effects on disk-like gold-nickel-platinum nanoswimmer's propulsion fuelled by hydrogen peroxide. Sensors and Actuators, B: Chemical. 2017;**239**:586

[74] Hu LX, Tao K, Miao JM, Grüber G. Hydrogen-peroxide-fuelled platinum-nickel-SU-8 microrocket with steerable propulsion using an eccentric nanoengine. RSC Advances. 2016;**6**: 102513

[75] Hu LX, Wang N, Miao JM, Grüber G. Investigation of temperature dependency on the propulsion of disk-like nanoswimmers. Procedia Technology. 2017;**6**:48

[76] Hu LX, Rehman S, Tao K, Miao JM. Characterization on three-dimensional trajectory of disk-like gold-nickel-platinum nanomotor using digital holographic imaging. ChemistrySelect. 2018;**3**:9634

[77] Hu LX, Tao K, Lim YD, Miao JM, Kim YJ. Self-steerable propulsion of disk-like micro-craft with dual off-center nanoengines. ACS Applied Energy Materials. 2019;**2**:1657

[78] Wang N, Kanhere E, Tao K, Hu LX, Wu J, Miao JM, et al. Investigation of a thin-film quasi-reference electrode fabricated by combined sputtering-evaporation approach. Electroanalysis. 2019;**31**:560

[79] Astorga SE, Hu LX, Marsili E, Huang YZ. Ordered micropillar array gold electrode increases electrochemical signature of early biofilm attachment. Materials & Design. 2020;**185**:108256

[80] Tao K, Yi HP, Yang Y, Chang HL, Wu J, Tang LH, et al. Origami-inspired electret-based triboelectric generator for biomechanical and ocean wave energy harvesting. Nano Energy. 2020;**67**: 104197